NEW TECHNOLOGIES IN THE OIL AND GAS INDUSTRY

Edited by **Jorge Salgado Gomes**

New Technologies in the Oil and Gas Industry
http://dx.doi.org/10.5772/3216
Edited by Jorge Salgado Gomes

Contributors

Alexandre Andrade Cerqueira, Monica Regina da Costa Marques, Adrián J. Acuña, Oscar H. Pucci, Graciela N. Pucci, Adesina Fadairo, Olugbenga Falode, Churchill Ako, Abiodun Adeyemi, Anthony Ameloko, Sergio Abreo, Carlos Fajardo, William Salamanca, Ana Ramirez, Airam Sausen, Paulo Sausen, Mauricio de Campos, Hugo Caetano, Freddy Humberto Escobar, Patrick W. M. Corbett, Mohammed Zubair Kalam

Published by InTech

Janeza Trdine 9, 51000 Rijeka, Croatia

Publishing Process Manager Tanja Skorupan
Technical Editor InTech Prepress, Novi Sad
Cover InTech Design team

Additional hard copies can be obtained from orders@intechopen.com

New Technologies in the Oil and Gas Industry, Edited by Jorge Salgado Gomes
p. cm.
ISBN 978-953-51-0825-2

For EU product safety concerns:
IN TECH d.o.o., Prolaz Marije Krucifikse Kozulić 3, 51000 Rijeka, Croatia,
info@intechopen.com or visit our website at intechopen.com.

Meet the editor

Jorge Salgado Gomes has 30 years of experience in the industry with several companies (BP Minerals, EDMA, PARTEX, QGPC, Hot-RC2 and ADCO) and academia (The Petroleum Institute). He holds a Master of Science in Geology from Oporto University (Portugal) and a MEng and PhD degrees in Petroleum Engineering from Heriot-Watt University (Edinburgh). During his career Dr. Gomes has worked in exploration and production and was also involved in large field development activities (major fields producing > 200,000 BOPD). Currently, Dr. Gomes is a PARTEX Chaired Professor at the Petroleum Institute in Abu Dhabi, doing research and teaching reservoir characterization. He is a member of SPE, AAPG, EAGE and SCA. He is the co-author of a university textbook about the oil and gas industry (ISBN 978-972-31-1228-3), and has published several papers.

Contents

Preface

This book seeks to cover a wide ground of research topics in different subjects of the oil and gas industry. The subject of new technologies is very broad, and the content of this book presents a small selection of some of them. Its content is intended for the use of geoscientists and petroleum engineers interested in reading the recent advances of certain oil and gas technologies.

Due to the large number of technologies available it was not possible to gather experts from all areas of technology. Therefore, a great amount of gratitude is given to the several oil and gas professionals, who kindly volunteered to share their valuable experiences and knowledge with all of us.

This book covers HSE (Chapters 1-3) to IT and Modeling (Chapters 4-6), Reservoir Management (Chapters 7-8) and Core Analysis (Chapter 9). Of these, Chapter 7 contains material that would be too detailed and advanced for some of the readers. Chapter 8 rounds off the book with a summary of the role of geosciences and engineering in field development. The chapters do not have to be read in numerical order and can be used individually.

The further technologies missing from this book must be acknowledged for further reference: Drilling Technologies, Reservoir Simulation, EOR, SCAL, PVT, Flow Assurance, Geomechanics, i-Field etc. but, we leave these for the second edition of the book.

I thank these authors on behalf of our industry for participating in this project and for allowing us to share the knowledge of research that is prevalent in furthering our understanding of a subject matter that is so close to us. I also hope that the release of this book will inspire others to work with us to release the second edition encompassing the remaining topics mentioned above.

I also want to thank Ms. Tanja Skorupan, InTech Publishing Process Manager, for her valuable support with the logistics, and for keeping me posted on the progress and time schedule of the book. If there are errors in this book, please forgive me, it is because I failed to detect and correct them.

Jorge Salgado Gomes
PARTEX Chair Professor,
The Petroleum Institute, Abu Dhabi,
United Arab Emirates

Section 1

HSE

Chapter 1

Electrolytic Treatment of Wastewater in the Oil Industry

Alexandre Andrade Cerqueira and Monica Regina da Costa Marques

Additional information is available at the end of the chapter

http://dx.doi.org/10.5772/50712

1. Introduction

Industrial development in recent decades has been a major contributor to the degradation of water quality, both through negligence in treatment of wastewater before discharge into receiving bodies and accidental pollutant spills in aquatic environments [1].

The importance of oil to society is unquestioned. It is not only a major source of energy used by mankind, but its refined products are the raw material for the manufacture of many consumer goods [2].

A world without the amenities and benefits offered by oil would require a total change of mindset and habits among the population, a total overhaul of the way our society works. At the same time, the oil industry is a major source of pollutants that degrade the environment, with the potential to affect it at all levels: air, water, soil, and consequently, all living beings on the planet [2].

Oil and its derivatives are the most important pollutants, due to, among other factors, the increasing amounts that have been extracted and processed. Also, carelessness and neglect of safety standards and routine maintenance of equipment (pipelines, terminals, platforms) aggravate the water pollution problems caused by the oil industry [3].

Due to the negative environmental impacts of exploration and production of oil, new more restrictive environmental laws and regulations have been issued. It is estimated that in the United States alone, the oil industry will need to invest about 160 billion dollars in actions to protect the environment over the next 20 years to meet environmental legislation more demanding than currently adopted in Brazil [4].

One of the crucial points to be attacked is the issue of water production, which is generated in this activity, which is increasing in volume as they get older wells and new wells are drilled [5]. On average for each m^3/day of oil produced, 3-4 m^3/day of water is produced, although this figure can reach up to 7 m^3/day or even more in exploration, drilling and production. The water produced along with oil corresponds to 98% of the effluents. It contains salts, oils and other toxic chemicals in addition to having high temperature and no oxygen [6].

According to [7], treatment of produced water is an urgent matter in view of the high daily volume. Different processes have been described for the treatment of such effluent, but the most frequently used are chemical destabilization [8,9] and electrochemical destabilization [10,11]. Biological processes are rarely used since these effluents usually contain biocides [12].

The use of EF can enable the release into receiving bodies or reinjection in wells of the treated effluent by reducing the organic load and removing oily and solid particles in suspension [13].

According to [14], the current EF technology inherently involves the formation of an impermeable oxide layer on the cathode and deterioration of the anode due to oxidation. This leads to loss of efficiency of the electroflocculation unit. These limitations of the process have been decreased to some extent by the addition of parallel plate electrodes in the cell configuration. However, the use of alternating current in EF retards the normal mechanisms of electrode deterioration that are inherent in DC system due to cyclical energization, thus increasing the electrode life.

In the present text, we evaluate the efficiency of electroflocculation with direct current and variable frequency alternating current with the use of aluminum electrodes for the treatment of oily wastewater from actual production.

1.1. Petroleum exploration and production

Petroleum is the name given to natural mixtures of hydrocarbons, which can be found in the solid, liquid or gaseous state depending on the conditions of temperature and pressure [15].

Oil is a combination of carbon and hydrogen molecules and is less dense than water, with a characteristic odor and color varying from black to brown. Although the subject of much discussion in the past, today oil's organic origin is accepted. Oil exploration and production is one of the most important industrial activities of modern society and its derivatives have many industrial applications. Because of the need to meet the growing demand for the product, the extraction of oil has increased greatly in recent decades. However, this extraction to meet world oil demand causes damage to the environment, with the main culprit being produced water [16,17].

1.2. Oily water

Oily water is a generic term used to describe all water which contains varying amounts of oils and greases in addition to a variety of other materials in suspension. These can include sand, clay and other materials, along with a range of dissolved colloidal substances, such as detergents, salts, metal ions, etc. To meet environmental standards for disposal and/or the characteristics necessary for reuse, the treatment of oily water can be complex, dependent on highly efficient processes.

In the petroleum industry, oily water occurs in the stages of production, transportation and refining, as well as during the use of derivatives. However, the production phase is the largest source of this pollution. During the production process, oil is commonly extracted along with water and gas. The associated water can reach 50% of the volume produced, or even approaching 100% at the end of the productive life of wells. The discharge or reinjection of this co-produced water is only permitted after removal of oil and suspended solids to acceptable levels [18].

The terms "produced water," "petroleum water", "formation water" and "oily water" are used to refer to the water extracted along with oil [17,19].

The composition of this produced water is very complex. Depending on its origin it can contain a wide variety of chemicals such as organic salts, aliphatic and aromatic hydrocarbons, oils and greases, metals, and occasionally radioactive materials. A striking feature of the water coming from offshore oil is its high salinity [17, 20, 21], which expressed as chloride ions (Cl^-) can reach 120 g/L [22].

In oil wells under the seabed, the amount of this wastewater can reach 90% of all effluent during the production of oil and can be 7-10 times higher than the oil extracted from a given well [17, 21].

A new oil field produces little oily water (about 5-15% of the total oil produced). However, as the well becomes exhausted, the water volume can increase significantly, to the range 75-90%. This excessive production of water has become a major concern in the oil and gas industry [23]. Before disposal into receiving bodies or use for re-injection into wells, it is necessary to treat this water because the large amounts of pollutants cannot be discharged into the marine environment [24].

1.3. Electrocoagulation

EC is a process that involves the generation of coagulants "in situ" from an electrode by the action of electric current applied to these electrodes. This generation of ions is followed by electrophoretic concentration of particles around the anode. The ions are attracted by the colloidal particles, neutralizing their charge and allowing their coagulation. The hydrogen gas released from the cathode interacts with the particles causing flocculation, allowing the unwanted material to rise and be removed (Figure 1). Various metals have been tested as electrodes, such as aluminum, iron, stainless steel and platinum [25].

The theory of EC has been discussed by several authors, and depending on the complexity of the phenomena involved can be summarized in three successive stages of operation:

a. Formation of a coagulating agent through the electrolytic oxidation of the sacrificial electrode, which neutralizes the surface charge, destabilizes the colloidal particles and breaks down emulsions (coagulation – EC step);
b. the particle agglutination promoted by the coagulating agent facilitates the formation and growth of flakes (flocculation – EF step) and,
c. generation of micro-bubbles of oxygen (O_2) at the anode and hydrogen (H_2) at the cathode, which rise to the surface and are adsorbed when colliding with the flakes, carrying the particles and impurities in suspension to the top and thereby promoting the clarification of the effluent (flotation – electroflotation step).

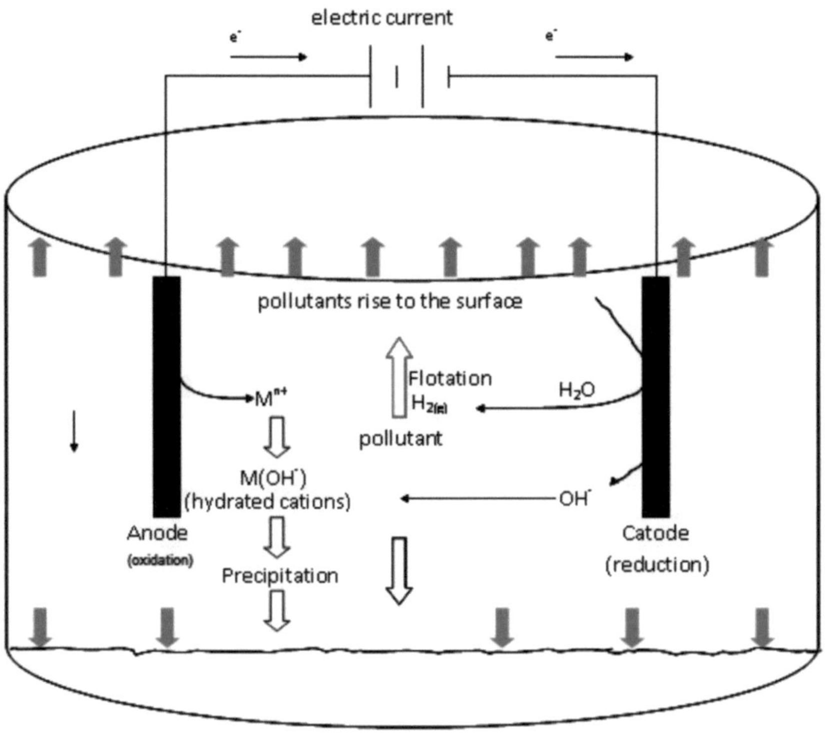

Figure 1. Schematic diagram of an electrocoagulation cell with two electrodes. Source: Adapted from [26].

Processes for electrochemical treatment of effluents have been described in the literature since 1903. In recent years interest has been growing, especially because of its simplicity of operation and application to treat various types of effluents from various sectors, such as domestic sewage [27], laundries [29], restaurants [30] steel mills [31], textile mills [32], and tanneries [33], facilitating the removal of metal ions [28], fluoride ion [34], boron [35] and oils [7, 36-41].

Several types of reactors have been proposed in the literature: monopolar, bipolar etc. But the most widely used is the monopolar reactor [14]. In its simplest form, a monopolar EF reactor is composed of an electrolytic cell with an anode and a cathode. In this case, large-area electrodes must be used, or electrodes connected in parallel. In the parallel arrangement, the electric current is divided among all the electrodes in relation to the resistance of individual cells. Thus, a lower potential difference is required in connection of this type when compared to a series arrangement.

For electrodes in series, a higher potential difference is required for a given current flow, because the electrodes are connected in series and have a higher resistance. The same current, however, runs through all the electrodes, and the current is divided among all the individual electrodes of the cells [14].

In the case of the bipolar reactor, the sacrifice electrodes are placed between the two electrodes in parallel (called conductive plates), without any electrical connection. Only two monopolar electrodes are connected to the power source, with no interconnection between the sacrifice electrodes. When the current passes through the two parallel electrodes, the neutral sides of the plate acquire an opposite charge than monopolar electrode. The external electrodes are monopolar and the internal ones are bipolar.

According to [42], most of the setups for treatment of effluents, the electrodes are made of identical material, mainly due to the following reasons:

- equal electrodes, made of the same material, have the same electrode potential;
- electrodes of different materials imply the use of materials other than iron or aluminum, which increases the cost;
- electrodes of the same material suffer the same wear, which simplifies their replacement.

In any electrochemical process, the electrode material has a significant effect on the effluent treatment. For the treatment of drinking water, it should be nontoxic, have low cost and be readily available [31].

Generally, however, iron electrodes have the disadvantage that the effluent has a pronounced green or yellow color during and after treatment. This coloration comes from the Fe^{2+} (green) and Fe^{3+} (yellow) generated in the electrolytic treatment. In contrast, with aluminum electrodes the final effluent is clear and stable, with no residual coloring.

In the work presented by [43], when aluminum and iron electrodes were tested under the same conditions, using direct current, the results for COD, turbidity and suspended solids were better for the aluminum than the iron electrodes. This advantage was also observed by [30]. However, when comparing the removal of arsenic by iron and aluminum electrodes, [31] found that the iron electrode was better because it showed 99% removal to 37% for aluminum. This difference was explained because the adsorption capacity of the $Al(OH)_3$ + by As^3 is much smaller than that of $Fe(OH)_3$.

Tests carried out by [44] of COD, phenols and turbidity of hydrocarbons from a petrochemical plant, using iron and aluminum electrodes, showed better performance by aluminum electrodes.

According to [45], just as in electrocoagulation, the removal of pollutants closely depends on the size of the bubbles generated, while energy consumption is related to the electrolytic cell design, electrode materials, arrangement of electrodes and operating conditions, such as current density, conductivity of the effluent and electrolysis time, among others. The difference in size of the bubbles in the effluent depends on the pH, current density, electrode material and surface condition of the electrodes.

The mechanism of EC is highly dependent on the chemistry of the aqueous medium, especially conductivity. Moreover, other characteristics such as pH, particle size and concentration of the constituents will influence the electrocoagulation process [14].

In an EC reactor, the rate of coagulant addition is determined by the kinetics of the electrodes. The reactions at the electrodes are heterogeneous and take place in the interfacial region between the electrodes and the solution. Since the reaction consists of electron transfer via an interface, this reaction will be influenced by the characteristics of this interface, such as the potential difference that is established in equilibrium and changes in potential across the interface in function of distance.

The potential of the electrolysis is strongly dependent on current density, effluent conductivity, distance between the electrodes and the surface condition of the electrodes.

1.3.1. Parameters influencing electrolytic processes

1.3.1.1. pH

The EC process performance is greatly influenced by the pH of the solution [46]. Considering only mononuclear speciation, the total aluminum present in solution (α) at a given pH value can be calculated (Figure 2). This distribution diagram shows the extent of hydrolysis, which depends on the total metal concentration and pH. As the pH increases, the dominant species changes, in this case from the Al^{3+} cation to the $Al(OH)_4^-$ ion not participate in the coagulation reactions and tend to remain in solution [47].

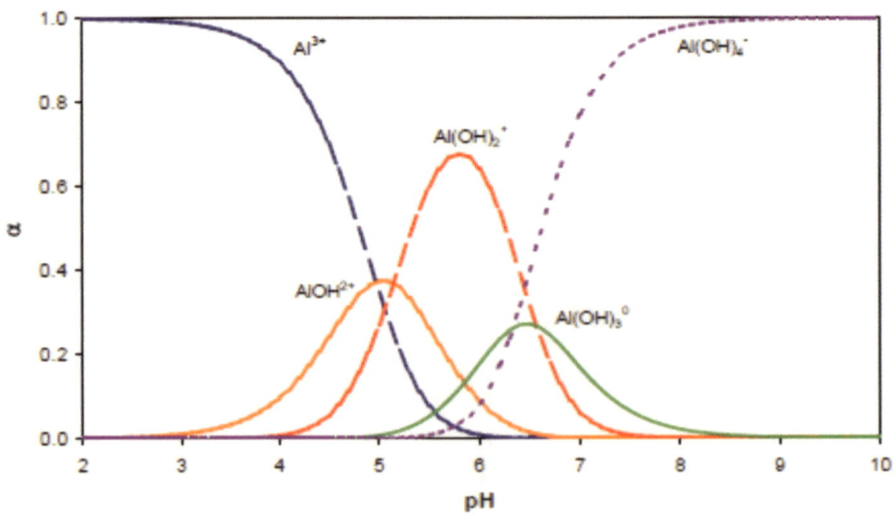

Figure 2. Diagram of distribution for Al-H_2O considering only mononuclear species (Source: [25]).

The solubility diagram for aluminum hydroxide, $Al(OH)_3$, is shown in Figure 3. The solubility boundary denotes the thermodynamic equilibrium that exists between the dominant aluminum species in solution at a given pH and the solid aluminum hydroxide. The minimum solubility (0.03 mg Al/L) occurs at a pH of 6.3 and increases as the solution becomes more alkaline or acid [48].

Thus, the active metal cations produced in the anode react with the OH^- ions produced at the cathode to form a metal hydroxide, which then acts as a coagulant with the polluting particles and the metal hydroxides, forming larger aggregates, which can both undergo sedimentation and be carried to the surface of the hydrogen bubbles generated at the cathode.

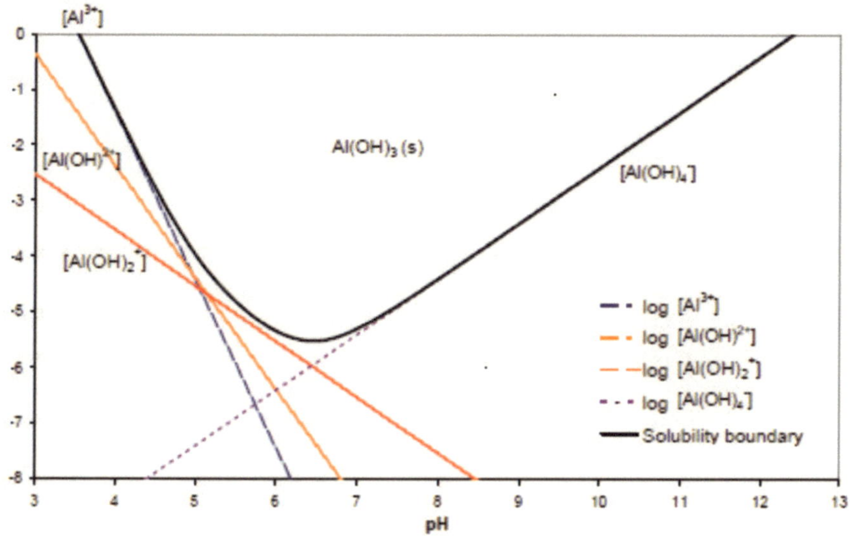

Figure 3. Solubility diagram of aluminum hydroxide $Al(OH)_3$ considering only mononuclear species of Al (Source: [25]).

The distribution and solubility diagrams presented above consider only mononuclear aluminum species, whereas in reality this system is considerably more complex. As the aluminum concentration increases, polynuclear complexes of aluminum can be formed and the aluminum hydroxide is precipitated, as illustrated by equation (1):

$$Al^{3+} \rightarrow Al(OH)^{3-n}_{n} \rightarrow Al_2(OH)^{4+}_{2} \rightarrow Al(OH)^{5+}_{4} \rightarrow Al_{13}\text{ complex} \rightarrow Al(OH)_3 \qquad (1)$$

By studying a continuous electrocoagulation process using aluminum electrodes and varying the pH, [7] observed that in acidic (pH <4) or alkaline (pH> 9) media, in which cationic or anionic monomeric species of aluminum are predominant, the emulsion remained stable and there was no decrease in COD. Moreover, at pH 5 to 9, the removal was 80%. Under these conditions (pH 5 to 9), the predominant species are polymeric complexes of aluminum and amorphous precipitate of aluminum hydroxide. The surface of the latter can be positively or negatively charged by adsorption of ions from the solution.

1.3.1.2. Distance between electrodes

The distance between the electrodes is an important variable to optimize operating costs. According to [49], when the effluent conductivity is relatively high, a bigger space between the electrodes should be used. In contrast, in situations of moderate value, it is recommended to use a smaller distance, as this will reduce power consumption without changing the degree of separation, because in this case the current would not be altered.

When testing a treatment system under the same electric current, [50] observed there was no difference in removal efficiency for different spacings between the electrodes. Therefore, the distance between them is considered to be only a factor for cost optimization.

In turn, [51] reported in their paper that with increasing distance between the electrodes, fewer interactions of ions of the solution with the coagulant will occur. The difference in conclusions between the two research teams can be attributed to a possible divergence in the conductivity value of each effluent, since if it was high in the first study (range between 100 and 140 mS cm^{-1}), there would not have been any change in removal efficiency, because even with a greater distance between the electrodes, there would be a minimum conductivity of the solution that would carry the current.

In the second study, the authors did not mention the effluent's conductivity value. However, it can be assumed it was lower than in the first study, because increasing the distance between the electrodes caused the interactions to decrease and there would need to be a minimum conductivity to ensure the transmission of electric current. Therefore, for there to be no difference in removal with changes in the spacing between electrodes, the treated solution must have a minimum electrical conductivity value.

The greater the distance between the electrodes, the greater must be the voltage applied, because the solution has resistance to the passage of electric current. Thus, according to the characteristics of the effluent, the distance between electrodes can be varied to maximize the efficiency of the process. For example, longer distances can be used when the effluent conductivity is relatively high, while the distance should be as small as possible when conductivity is low so that does not overly increase the need for power.

1.3.1.3. Electrical conductivity of the effluent

The increase in conductivity by addition of sodium chloride is known to reduce the cell voltage due to the reduction of the ohmic resistance of the effluent [43, 52]. Chloride ions can significantly reduce the adverse effects of other anions such as HCO_3^- and SO_4^{2-}.

The electrical conductivity of the effluent is a variable that affects the current efficiency, the cell voltage and power consumption. It is also important when optimizing the parameters of the system, since high conductivity associated with a small distance between electrodes minimizes the consumption of energy, but does not affect the efficiency of removing contaminants, as shown in [53].

When the electric conductivity of an effluent is too low, sodium chloride (NaCl) can be added to increase in the number of ions in solution. But this leads to oxidation of chloride ions to chlorine gas and OCl ions, which are strong oxidants capable of oxidizing organic molecules present in the effluent [54].

The reactions (2, 3 and 4) are [29]:

$$Cl^{-}_{(aq)} \rightarrow Cl_{2(g)} + 2e^{-} \quad (2)$$

$$Cl_{2(g)} + H_2O \rightarrow HOCl_{(l)} + H^{+}_{aq} + Cl^{-}_{(aq)} \quad (3)$$

$$HOCl_{(l)} \rightarrow H^{+}_{aq} + OCl^{-}_{(aq)} \quad (4)$$

According to [55], the power consumption does not diminish significantly when the conductivity of the solution is greater than 1.5 mS/cm.

The conductivity of the effluent, namely the capacity to conduct electrical current, should be directly proportional to the amount of ions present in the conductive liquid. These ions are responsible for conducting the electrical current. It is evident, then, that the higher the concentration of these ions in the effluent, the greater its ability to conduct electrical current and the greater the possibility of reactions between the substances present in the effluent, a positive factor which enables reduction of energy consumption.

1.3.1.4. Temperature

According to [56], the effect of temperature has as yet been little investigated in the electrocoagulation process. Some studies have shown that the efficiency achieved with aluminum electrodes increases with temperature up to the 60 °C, above which the efficiency decreases. However, the conductivity increases with increasing temperature, decreasing the resistivity and electric power consumption. Increasing the temperature of the solution contributes to increase the efficiency of removal, caused by the increase of the movement of the ions produced, facilitating their collision with the coagulant formed [57, 58].

1.4. Electrolytic processes applied to the treatment of oily wastewater

According to [36] in the 1980s, Zhdanov used iron and aluminum electrodes to break down emulsions and promote flocculation of wastewater impurities from drilling platforms, aiming at its reuse.

The EPA (1993) conducted studies on the use of innovative technologies for treating hazardous waste, using the technique of electrocoagulation with alternating current. The resulting apparatus was called the ACE Separator™. This technology introduces low concentrations of nontoxic aluminum hydroxide in the medium.

The effluents were prepared in order to reproduce the natural leakage to the underground reservoirs in soil washing operations. The main objective of these tests was to obtain optimal conditions for breaking oil-in-water emulsions and achieve reductions of soluble solids and loads of metal pollutants.

Experiments were carried out using a monopolar electrode of aluminum and the effluent used was prepared with 1.5% of diesel oil, 0.1% surfactant, 10 to 100 mg L of metals (Cu, Cd and Cr) and 3% soil containing 50% clay. Assays were performed at pH 5, 7 and 9. NaCl was added in the range from 1200 to 1500 mg/L to simulate salinity values found in contaminated media. The optimum operational conditions were: 4 A (ampere); space between the electrodes of 0.5 cm, duration of 3 to 5 minutes and frequency of 10 Hz. The pollutant removal efficiencies were 98% of TSS, 95% for TOC, 72% for Cu, 92% for Cr and 70% for Cd. However, fouling was observed on the plates of the electrodes.

Another research team [36] used the electrolytic process with titanium electrodes to promote the oxidation of pollutants in the oil industry. The process was tested with effluents of low and high salinity containing sulfides, ammonia and phenol, besides organic matter. Studies were also performed simulating the generation of chlorine by electrolysis with salinity levels similar to those found in the effluent.

The results demonstrated the possibility of using the electrolyte process in both situations, but that it was particularly advantageous when used in high salinity effluent due to the high conductivity, which allows oxidation with lower energy consumption.

[39] conducted experiments with electrolysis to remove the COD, O&G and turbidity from olive oil residues in the presence of H_2O_2 and a flocculating agent generated "in situ" via iron and aluminum electrodes. The iron electrode was more effective than aluminum. The COD removal efficiency was around 62-86%, whereas the removal of O&G, and turbidity was 100%. The current density ranged from 20-75 mA cm^2, depending on the concentration of H_2O_2 and coagulating agents. Using petrochemical industry effluents, [59] conducted tests of chemical coagulation (jar test) and electrocoagulation on a laboratory scale. The tests allowed comparing the removal efficiencies of organic matter by electrocoagulation and chemical coagulation, and comparing the efficiencies of these treatments in laboratory scale with those obtained in the stage of physical-chemical treatment (chemical coagulation and flocculation). In all cases the efficiency of removal of organic matter were evaluated according to reduction of COD.

In the chemical coagulation assays, the authors used aluminum sulfate. The parameters evaluated were the optimum pH for coagulation and optimum coagulant dosage. Tests were carried out of the electrolytic process in batch with aluminum electrodes. The parameters analyzed were: temperature, applied potential, initial pH, distance between electrodes, number of electrodes and electrode wear. The efficiency of the electrocoagulation process showed values up to three times higher than the monthly average obtained by the petrochemical industry using chemical coagulation and flocculation.

Another research team [13] studied the possibility of applying electrocoagulation to treat synthetic wastewater in the oil industry. This effluent was prepared in the proportion of 33 L of water to 50 ml of crude oil in a vessel with mechanical stirring for 30 minutes. A single-compartment electrochemical cell was used to generate bubbles, operating in continuous system with power feed at the top and treated effluent outlet at the bottom. The anode was made from titanium, called a DSA® and the cathode material was grade 316 stainless steel.

Electrolysis was carried out using current density of 20 mA/cm^2, flow rates of 800 and 1200 mL/h^{-1} and electrolysis times 150 and 180 minutes, respectively. The results showed that it was possible to obtain removal of COD and O&G greater than 90%.

Other tests have been performed to investigate the treatment of oily wastewater from washing the holds of ships, using the technique of electrocoagulation. The process was evaluated in laboratory scale and involved the use of two types of electrodes (iron and aluminum). The results showed that the best performance was obtained using the iron electrode [60].

The system operated with current of up to 1.5 A, for 60 and 90 minutes. The removal rates of BOD and O&G were 93 and 96%, respectively, while the COD removal rates were 61 and 78%, depending on the treatment time. Finally, 99% of the hydrocarbons were removed. Electrocoagulation was also effective for clarification of the effluent. Removal rates of 99 and 98% were measured for TSS and turbidity, respectively.

To verify the efficiency of treatment by synthetic cutting fluid, electroflotation was performed to characterize the fluid before and after treatment. The parameters analyzed were: pH, turbidity, metals, total phosphorus, COD, BOD and O&G. The results were quite satisfactory. The EF efficiency showed partial removal of contaminants in the cutting fluid, but the concentration of O&G exceeded the maximum limit for disposal according to relevant legislation [61].

Using the EC technique with a perforated aluminum electrode to separate oil emulsified in water, [62] found that the perforated electrode facilitates the passage and upward movement of the oil droplets to the surface. The authors observed that at 5 V and 0.4 A, the oil removal efficiency was 90% at pH 4.7 during 30 min of electrolysis, and the optimal salt concentration was 4 mg / L. The oil removal rate from the effluent increased with decreasing pH and lower salinity.

With the goal of removing Cu^{+2}, Zn^{+2}, phenol and BTEX from the produced water, [63] studied two types of electrochemical reactors: one using electroflocculation and the other electroflotation. In the former, an electric potential was applied to a solution containing NaCl, through electrodes of Fe, which with the dissolution of the metal ions generated Fe^{+2} and gases. An appropriate pH, these gases caused coagulation/flocculation reactions, removing Cu^{+2} and Zn^{+2}. In the second setup, a carbon steel cathode and DSA®. of DSA® Ti/TiO_2-RuO_2-SnO_2 were used, in a solution containing NaCl, which produced strong oxidizers such as HOCl and Cl_2. These promoted degradation of BTEX and phenol at different flow rates. The Zn^{+2} was removed by electrodeposition or by the formation of $Zn(OH)_2$ due to the increased pH.

Assessing the removal of oil from a synthetic emulsion by the electrocoagulation-flotation process, [45] observed the influence of operating parameters on the rate of reduction of COD, initial oil concentration, current density, electrode separation, pH and electrolyte concentration. NaCl was added to increase the solution's conductivity. The initial pH of the emulsion was 8.7. The Zeta potential had an average value of -75 mV, indicating emulsion stability. The author found that the best conditions for removal were current density of 4.44 mA/cm^2, treatment time of 75 min, distance between electrodes of 10 mm and concentration of the electrolyte (NaCl) of 3 g/L.

Studying the treatment of a synthetic effluent and a real produced water sample for removal of oil by the Fenton process, electroflotation and a combination these two, [64] first evaluated the Fenton and electroflotation processes individually and optimized the parameters for evaluating the combined process. The Fenton process, using Fe^{+2} and H_2O_2, obtained a peak oil removal of around 95% after 150 minutes and 50% removal after 57 minutes. The EC with the optimized volt (V) value managed to remove 98% of the oil after 40 minutes. The combined process using the optimized parameters for each process achieved removal of 98% after 10 minutes and 50% after 1 to 3 minutes. The combined process proved to be much more efficient than the procedures alone.

[65] evaluated the removal of sulfate and COD from oil refinery wastewater through three types of electrodes: aluminum, stainless steel and iron. They investigated the effects of current density, electrode array, electrolysis time, initial pH and temperature for two samples of wastewater with different concentrations of COD and sulfate. The experimental results showed that the aluminum anode and cathode was more efficient in the reduction of both contaminants. The results demonstrated the technical feasibility of electrocoagulation as a reliable method for pre-treatment of contaminated wastewater from refineries.

[66] in their experiments showed that treatment of synthetic wastewater emulsified water produced by EF, produced better results when used at a frequency of 60 Hz alternating current, initial pH 9, electrolysis time of 3 minutes and application of intensity current of 3 A. The results of tests on simulated wastewater produced water resulted in high removal efficiencies of organic load reaching 99% removal of oil and grease, color and turbidity. Compared to the flocculation trials using *Jar-Test*, the EF demonstrated highly efficient for the treatment of effluent water production in order to remove oil and grease emulsions, color and turbidity with no addition of chemical reagents or pH adjustment. *Jar-Test* trials were not effective consume high amounts of aluminum sulfate and low efficiency of removal of parameters. The main advantage of alternating current electrolysis in comparison with the direct current is less wear of the electrode mass. By using the same assay conditions for both technologies in 60 minutes oxidized alternating current of 1.6 g Al electrode while the oxidized direct current electrode of 3.4 g Al.

2. Variable frequency AC electroflocculation

In this work, we used variable frequency electroflocculation, which consists of using alternating current from the power grid at 60 Hertz and varying the voltage and frequency between 1 Hz to 120 Hz. This alternating current was generated by reconstituting the sinusoidal form of the input current in a conversion system with vector control, which generates a pulse-controlled formation time (period) adjusted by a programmable base time through a system of microprocessors.

This system triggers an oscillator to form a new waveform that has a peak residence time large enough to have conduction at a given polarity. The evaluation of EF with alternating current for the treatment of effluents from oil platforms can be of great importance to develop treatment processes that are fast, efficient and cost-effective. The aim of this experiment was to develop and evaluate in laboratory scale a variable frequency electroflocculation (EF) system for the treatment of oily water generated in offshore oil production and to compare the results against those produced using direct current.

2.1. Electroflocculation units

AC and DC electroflocculation units consisted of a glass electrolytic cell with capacity of 1 liter under magnetic stirring, in which an electrode was inserted vertically (monopolar in parallel) in a honeycomb arrangement, made of seven interspersed aluminum plates. These plates measured 10 cm long, 5 cm wide and 3 mm thick and were separated by spacers of 1 cm each. After a predetermined electrolysis period, we waited for 30 minutes for complete flotation of the emulsion to occur. Through a tap at the bottom of the beaker, the treated effluent was removed to assess the efficiency of electroflocculation (EF), which was done by monitoring, in triplicate, the following parameters: pH, conductivity, turbidity and color.

2.1.1. DC unit

The DC electroflocculation unit used a voltage of up to 15 V. First AC power (110/220 V) was applied to a potentiometer connected to a step-down transformer, feeding the secondary stage rectifier bridge responsible for providing DC power to the electrodes by a polarity reversing switch, connected to a meter showing voltage (V) and current (A). These readings guide the operator regarding the parameters of honeycomb electrode array. Figure 4 shows the diagram of the experimental DC setup.

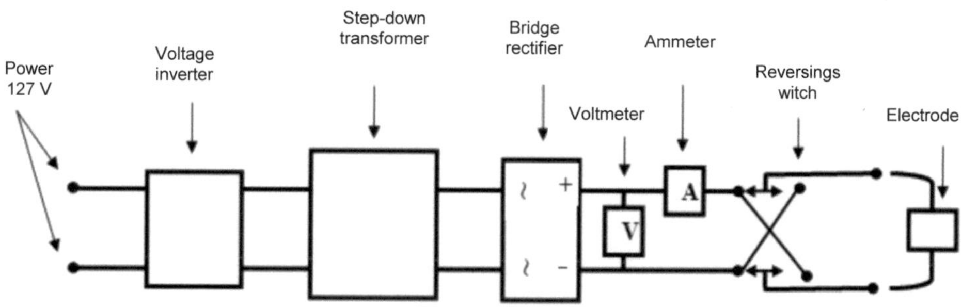

Figure 4. Schematic diagram of the experimental DC unit.

2.1.2. Variable frequency AC unit

The alternating current at a voltage of up to 15 V and variable frequency between 1 Hz and 120 Hz was obtained from a Weg CFW0800 AC/AC converter and a step-down transformer

(Tecnopeltron PLTN model 100/15). In this setup, the input power at 60 Hz from the grid is converted to variable frequency output of 1 to120 Hz in order to obtain AC power at the desired level. As with the DC setup, there is a meter to indicate the voltage (V) and current (a), to guide the operator.

Figure 5 shows a block diagram where the 60 Hz current from the grid feeds a frequency converter with variable output from 1 Hz to 120 Hz, connected to a variable voltage step-down transformer, thereby providing appropriate frequency and voltage to the electrode. In the rectification step that occurs in the variable frequency converter, the power is transformed into DC. Then the new direct current is treated in the oscillator module which converts it into pulses with controlled width, forming a new AC waveform, with a frequency that can vary between 1 Hz and 120 Hz depending on the level of feedback (reference) from the load controller. Thus, it has a sinusoidal waveform where the period varies with the load, to obtain the best performance at active power levels.

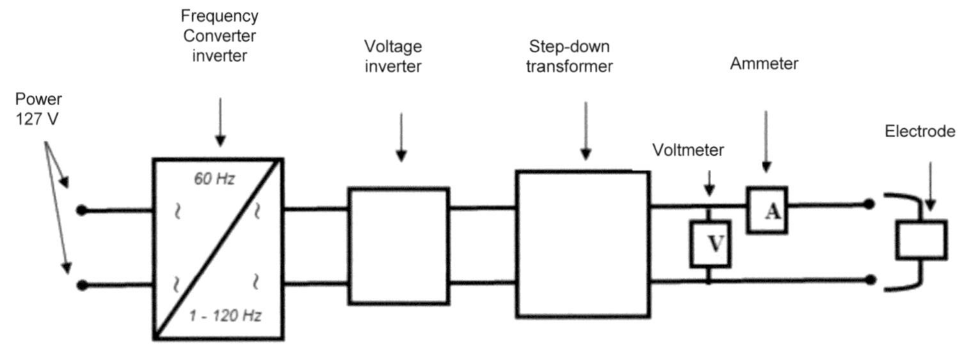

Figure 5. Schematic diagram of the experimental AC unit

The electrode is the central element for treatment. Thus, the proper selection of its materials is very important. The most common electrode materials for electroflocculation are aluminum and iron, since they are inexpensive, readily available and highly effective. In this experiment we used a hive array of seven interspersed aluminum plates measuring 10 cm long, 5 cm wide and 3 mm thick. The plates were separated by spacers (1 cm thick each), allowing varying the distance between the electrodes.

The electrodes were connected to specific instruments to control and monitor the current and voltage applied to the system, namely a frequency converter/regulator, potentiometer, step-down transformer, voltmeter, ammeter, bridge rectifier and polarity reversing switch.

Figure 6 shows an example of hive aluminum electrodes.

Figure 6. Hive aluminum electrodes containing eigth plates.

2.2. Tests with real effluent using oily water from the oil industry

Through laboratory tests we noted that the real effluent yielded obtained from an oil company did not have high salinity and had high oil and grease content (60 g / L), so it was not characterized as produced water but rather as oily effluent.

We conducted there tests:

1. the first using the effluent as received;
2. the second adding 60g / L of natural salt; and
3. the third adding salt 60g / L of salt plus emulsifiers.

Table 1 shows the values obtained with the AC and DC electroflocculation processes with the original effluent as received. The results of the AC setup were obtained with the maximum current of the unit (i = 2.5 A), due to the low salinity. The voltage was 11 V. In the case of direct current, the unit only reached a maximum of 1.6 A, so we added 1 g of salt to obtain the same current intensity as the AC unit. By adding salt to the effluent in the DC system, there is an improvement in removal efficiency.

The analysis of the oil and grease parameter (supernatant) was carried out separately, while the remaining parameters were analyzed with the subnatant phase of the effluent. The AC and DC electroflocculation tests were performed with the effluent containing an oil and grease content of 60g/L.

There was an increase in pH during the final AC and DC tests, attributed to the generation of OH-ions during the water reduction step.

$$2H_2O_{(l)} + 2e^- \rightarrow H_{2(g)} + 2OH^-_{(aq)} \quad (5)$$

Although these ions are also used to form the coagulating agent, the remaining quantity results in an increase of the pH value. This was also observed by [43, 67, 68].

Parameters	Oily wastewater	AC	DC
pH	6.7	8.3	8.2
Turbidity (NTU)	840	2	2
Color (Abs. 400 nm)	0.46	0.02	0.01
Salinity (mg/L)	279	340	1210
Conductivity (μS/cm)	580	702	2238
TDS (mg/L)	408	498	1680
Phenols (mg/L)	0.5	< 0.1	< 0.1
Sulfides (mg/L)	2.8	< 0.5	< 0.5
Ammonia nitrogen (mg/L)	36.0	3.7	1.5
O&G (mg/L)	60000	18	22
Current (A)	0	2.5	2.5
Tension (V)	0	11.0	8.5

Legend: Data: time = 15 min., $D_{electrode.}$ = 1 cm, vol. = 3L, i = 3 A, f = 60 Hz AC. Note: In the DC setup, 1 g of salt was added to increase the conductivity and the current intensity of the equipment, because it was unable to reach the same as with the AC setup.

Table 1. Data from the EF treatment process.

Despite the low conductivity of the effluent, there was high removal of pollutants and complete clarification after treatment for 15 minutes. The removal of ammonia, phenols and sulfides may have been obtained by drag of the gas phase (electroflotation).

Figure 7 shows the evolution of the tests of the raw effluent and after treatment with AC and DC.

Figure 7. EF tests with real effluent with low salinity. Legend: a) raw effluent; b) effluent treated with AC electroflocculation; c) effluent treated with DC electroflocculation; and d) comparison of raw effluent with samples treated with AC and DC electroflocculation. Note: The tests performed with t = 15 min.

The test results using the effluent plus NaCl to bring the salinity to 60 g / L, to simulate produced water, are shown in Table 2.

Parâmeters	Oily wastewater	AC	DC
pH	5.9	6.3	6.3
Turbidity (NTU)	1000	10	16
Color (Abs.)	0.63	0.03	0.05
Salinity (g/L)	50.8	46.0	47.2
Conductivity (mS/cm)	91.5	72.4	75.3
TDS (g/L)	64.9	59.6	59.6
Phenols (mg/L)	0.5	< 0.1	< 0.1
Sulfide (mg/L)	2.8	< 0.5	< 0.5
Ammonia nitrogen (mg/L)	36.0	2.9	2.7
O&G (mg/L)	60000	30	35
Mass electrode (g)	-	0.16	0.23
Current (A)	-	2.5	2.5
Tension (V)	-	1.5	2.0

Data: vol. = 3 L, t = 6 min., $D_{electrode.}$ = 1 cm, vol. = 3 L, i = 3 A, f = 60 Hz AC.

Table 2. Results obtained during tests of AC and DC electroflocculation using effluent with high salinity.

Since the electrolytic process involves corrosion of the electrode, according to Faraday's laws, there is mass loss of the electrodes. We measured this electrode mass loss by the weight difference before and after each test. High removal efficiencies were observed in tests with the high-salinity effluent. Electrode mass consumption with alternating current was 33% lower than with direct current, with all other conditions the same. Low voltage was applied during electrolysis in the tests to achieve high pollutant removal efficiency. In previous tests, the low conductivity greatly increased the voltage required by the system, leading to high energy consumption. This indicates that high conductivity greatly favors the electrolytic process.

Figure 8 shows the evolution of the tests with high-salinity effluent using the EF technology with alternating current and direct current. Note the total clarification after the tests without filtration.

Table 3 shows the results of the high-salinity effluent treated with the AC and DC processes.

As can be seen from the above table, the high turbidity, color and O&G were almost completely removed (above 99%). The voltage applied to the electrodes was very low, resulting in high efficiency of the technique. The electrode mass consumed with DC was 31% higher than with AC.

A hypothesis for the lower electrode consumption with alternating current is that since DC only flows in one direction, there may be irregular wear on the plates due to the onslaught of the current and subsequent oxidation occurring in the same preferential points of the electrode. In the case of AC, the cyclical energization retards the normal mechanisms of attack on an electrode and makes this attack more uniform, thus ensures longer electrode life.

Figure 8. EF tests with high-salinity effluent. Legend: 60g / L NaCl: a) raw wastewater containing 60g / L of O&G; b) raw wastewater being mixed salt, for EF testing; c) effluent treated with AC (left) and DC (right) electroflocculation.

Parâmeters	Oily wastewater	AC	DC
pH	6.4	6.7	6.7
Turbidity (NTU)	11050	8	10
Color (Abs.)	7.57	0.03	0.04
Salinity (g/L)	50.8	47.5	47.8
Conductivity (mS/cm)	91.5	87.4	87.0
TDS (g/L)	64.9	61.9	61.5
Phenols (mg/L)	0.5	0.2	0.2
Sulfide (H_2S) (mg/L)	2.8	1.2	1.0
Ammonia nitrogen (mg/L)	36	4	5
O&G (mg/L)	60000	32	30
Mass electrode (g)	-	0.18	0.26
Current (A)	-	2.5	2.5
Tension (V)	-	1.5	2.0

Note: Data: vol. = 3 L, t = 6 min.

Table 3. Results of treating high-salinity effluent with AC and DC electrolysis.

Figure 9 shows the evolution of EF treatment of high salinity effluent during the stages of development using AC and DC.

Figure 9. Tests of EF with high-salinity effluent. Legend: (60g/L) of salt. Processing steps: a) emulsified effluent, b) effluent undergoing electroflocculation with formation of supernatant sludge, c) treated effluent, with sludge formation; d) original effluent and after treatment with AC (left) and DC (right) electroflocculation.

3. Conclusions

In the present study, we could confirm that the EF process produces satisfactory results for treatment of oily wastewater, allowing its discharge into water bodies or reinjection in oil formations. The AC technology was highly effective, both with the original oily water as received and with the simulated produced water after addition of salt.

Overall, the results confirm the potential of the technique, which through simple and compact equipment, can be employed for the decontamination of organic compounds. The results of tests on oily water resulted in high organic load removal efficiencies, reaching 99% removal of oil and grease, color and turbidity, along with high removal of phenols, ammonia and sulfides.

The biggest advantage of AC versus DC electroflocculation is the lower electrode wear with the former technique. When using the same testing conditions and time of 6 minutes for both technologies, the efficiency was above 30%. The AC electroflocculation technique seems to be a promising alternative in the treatment of oily wastewater from the oil industry.

4. Acronyms

A - Ampere
AC - Alternate Current
ACE Separator™ - Alternating Current electrocoagulation
BOD - Biochemical Oxygen Demand
BTEX - Benzeno, Tolueno, Etil-Benzeno and Xilenos)
COD - Chemical Oxygen Demand
DC - Direct Current
$D_{electrode}$ – Distance between electrodes
DSA® - Dimensionally Stable Anode
EC - Electrocoagulation
EF - Electroflocculation
EPA - Environmental Protection Agency
Hz - Hertz
O&G - Oils and Greases
pH_i – initial pH
TOC - Total Organic Carbon
TSS - Total Suspended Solids
V - Volt

Acknowledgements

The National Councel of Technological and Scientific Development (CNPQ). When National Institute of Oil and Grease (INOG), we thank the Environmental Technology Laboratory (LABTAM) for the tests performed during this study, the Graduate Program in Environmental Protection (PPG-MA/UERJ) for support, and Rio de Janeiro State Research Foundation (FAPERJ) for financing through a PhD scholarship.

Author details

Alexandre Andrade Cerqueira and Monica Regina da Costa Marques
Instituto de Química – Laboratório de Tecnologia Ambiental (LABTAM),
Programa de Pós-Graduação em Meio Ambiente (PPG-MA),
Universidade do Estado do Rio de Janeiro (UERJ), Brazil

Appendix

Components of the AC and DC units

a. Frequency converter/regulator

Frequency converters, also known as frequency inverters, are electronic devices that convert alternating voltage into direct current and then back into alternating current at variable voltage and frequency. In our setup, we used this device to convert the current from the grid frequency of 60 Hz into the frequency applied to the electrode, of 1 to 120 Hz.

b. Potentiometer

A potentiometer is an electronic component that has adjustable resistance, providing variable the output voltage to the electrode.

c. Step-down transformer

A step-down transformer is used to lower single-phase, two-phase or three-phase voltage, to adapt the grid voltage to the needs the equipment to be used. Its primary winding is electrically isolated from the secondary winding. A mesh (electrostatic shield) can be placed between the windings, which once grounded helps eliminate noise from the power grid. In the case of electrolytic treatment here, the input voltage of 110/220 V was transformed to output voltage according to the operational need, e.g., 15 V.

d. Voltmeter

This is a device that measures voltage in a circuit. The voltage is generally shown by a movable pointer or a digital display. They are designed so that the high internal resistance introduces minimal changes to the circuit being monitored. To measure the voltage difference between two points in a circuit, the voltmeter is placed in parallel with the section of the circuit between these two points. Therefore, to obtain precise measurements, the meter should have a very large resistance compared to the circuit. Voltmeters can measure DC and AC voltages, depending on the quality of the device.

e. Ammeter

An ammeter is an instrument used to measure the intensity of electrical current flow passing through the cross section of a conductor. Since the current flows through the conductor and the devices attached to it, to measure the current passing through a region of a circuit, the meter must be placed in series with the circuit section. Therefore, to obtain precise measurements, the ammeter should have a resistance that is very small compared to the circuit. Ammeters can measure continuous or alternating current.

f. Bridge rectifier

This is a device that converts alternating current (AC) (usually sinusoidal) into direct current.

g. Polarity reversal switch

This is an important element of treatment systems, by increasing the lifetime of the electrode and reducing the resistance of the system. It changes the anode to the cathode and back again at regular intervals, to reduce erosion and scaling.

5. References

[1] Limons L S (2008) Avaliação do potencial de utilização da macrófita aquática seca salvinia sp. No tratamento de efluentes de fecularia. 87f. Dissertação (Mestrado) - Universidade Estadual do Oeste do Paraná, Cascavel, Brazil.

[2] Mariano J B (2005) Impactos ambientais do refino de petróleo. Rio de Janeiro: Interciência.

[3] Magossi L R, Bonacella P H (1999) A poluição das águas. 17 ed. São Paulo: Ed. Moderna. (Coleção Desafios)

[4] Barboza, J. Impactos ambientais do refino de Petróleo. Rio de Janeiro: Interciência, 2005.

[5] Campos A L O et al. (2005) Produção mais limpa na industria de petróleo: o caso da água produzida no campo de Carmópolis/SE. In: Congresso Brasileiro de Engenharia Sanitária e Ambiental, 23, Campo Grande, MS. Anais, ABES.

[6] Thomas J E (2004) Fundamentos de Engenharia de Petróleo. 2. ed. Rio de Janeiro: Interciência.

[7] Cañizares P et al. (2007) Break-up of oil-in-water emulsions by electrochemical techniques. J. Haz. Materials, v. 145, n. 1/2, p. 233-240.

[8] Pinotti A, Zaritzky N (2001) Effect of aluminium sulphate and cationic polyelectrolytes on the destabilization of emulsified waste. Waste Manage, v. 21, p. 535 – 542.

[9] Rios G, Pazos C, Coca J (1998) Destabilization of cutting oil emulsions using inorganic salts as coagulants. Colloid Surf. v. A 138, p. 383–389.

[10] Carmona M ET AL. (2006) A simple model to predict the removal of oil suspensions from water using the electrocoagulation technique. Chem. Eng. Science, v. 61, p. 1233 – 1242.

[11] Calvo L S, et al. (2003) An electrocoagulation unit for the purification of soluble oil wastes of high COD, Environmental Program. v. 22, p. 57–65.

[12] Kim B R et al. (1992) Aerobic treatment of metal-cutting fluids wastewater. W. Env. Research, v. 64, p. 216–222.

[13] Santos A C et al. (2007) Tratamento de efluentes sintéticos da indústria do petróleo utilizando o método de eletroflotação. In: PDPETRO, 4, 2007, Campinas. Anais. Campinas, ABPG.

[14] Mollah M Y A et al. (2001) Electrocoagulation (EC): science and applications. J. Haz. Materials, v. 84, p. 29-41.

[15] Rosa A J (2006) Carvalho, R. S.; Xavier, J. A. D. Engenharia de reservatórios de petróleo. Rio de Janeiro: Interciência.

[16] Macedo V A P (2009) Tratamento de água de produção de petróleo através de membranas e processos oxidativos avançados. 92f. Dissertação (Mestrado) - Universidade de São Paulo, Lorena, Brazil.

[17] Oliveira E, Santelli R, Casella R (2005) Direct determination of lead in produced waters from petroleum exploration by electrothermal atomic absorption spectrometry X-ray fluorence using Ir-W permanent modifier combined with hydrofluoric acid. Analyt. Chim. Acta, v. 545, p. 85-91.

[18] Ramalho J B V S (1992) Curso básico de processamento de petróleo: tratamento de água oleosa. Rio de Janeiro: RPSE:DIROL:SEPET.

[19] Gabardo I T (2007). Tese de Doutorado, Universidade Federal do Rio Grande do Norte, Brazil.

[20] Henderson S B et al. (1999) Potencial impact of production chemicals on the toxicity of produced water discharges from North Sea Oil Platforms. M. Pollut. Bulletin, v. 38, n. 12, p.1141-1151.

[21] Dórea H et al. (2007) Analysis of BTEX, PAHs and metals in the oilfield produced water in the state of Sergipe, Brazil. Microchemical Journal, v. 85, p. 234-238.

[22] Campos J C et al. (2002) Oilfield wastewater treatment by combined microfiltration and biological processes. Water Research, v. 36, p.95-104.

[23] Vieira D S, Cammarota M C, Camporese E F S (2003) Redução de contaminantes presentes na água de produção de petróleo. In: Congresso Brasileiro de P&D em Petróleo e Gás, 2, Rio de Janeiro. Anais. Rio de Janeiro: Instituto Brasileiro de Petróleo e Gás.

[24] Cunha G M A et al. (2005) Tratamento de águas produzidas em campos de petróleo: estudo de caso da estação de Guamaré/RN. In: Congresso Brasileiro de Engenharia Química em Iniciação Científica, 6, Campinas. Anais. Campinas, Unicamp.

[25] Holt P (2002) Electrocoagulation unravelling and synthesising the mechanisms behind a water treatment process. 229f. Thesis (Ph. D.) - University of Sydney, Austrália.

[26] Mollah M Y A et al. (2004) Fundamentals, present and future perspectives of electrocoagulation, J. Haz. Materials, v. 114, p. 199-210.

[27] Wiendl W G (1998) Processos eletrolíticos no tratamento de esgotos sanitários. Rio de Janeiro: ABES.

[28] Torem M L et al. (2002) Remoção de metais tóxicos e pesados por eletroflotação. Saneamento Ambiental, v. 85, p. 46-51.

[29] Ge J, Qu J, Lei P, Liu J (2004) New bipolar electrocoagulation-electroflotation process for the treatment of laundry wastewater. Sep. Purif. Technology, v. 36, p. 33-39.

[30] Chen G, Chen X, Yue P L (2000) Electrocoagulation and electroflotation of Restaurant Wastewater. Journal Environmental Engineering. v. 126, n. 9, p. 858-863.

[31] Kumar P R, Chaudhari S, Khilar K C, Mahajan S P (2004). Chemosphere, 9, 55.

[32] Cerqueira A A, Russo C, Marques M R C (2009) Electroflocculation for textile wastewater treatment. Braz. J. Chem. Engineering, v. 26. p. 659-668.

[33] Murugananthan M, Bhaskar G, Prabhakar S (2004) Removal of sulfide, sulfate and sulfite ions by electrocoagulation. J. Haz. Materials, v. 109, p. 37–44.
[34] Hu C Y et al. (2005) Removal of fluoride from semiconductor wastewater by electrocoagulation-flotation. W. Research, v. 39, n. 5, p. 895 – 901.
[35] Yilmaz A E, Boncukcuoglu R, Kocakerin M M, Keskinler B (2005). J. Haz. Mater., 125, 1.
[36] Queiroz M S, Souza A D, Abreu E S V, Gomes N T, Neto O A A (1996) Aplicação do Processo Eletrolítico ao Tratamento de Água de Produção, CENPES-DITER-SEBIO, RT: Rio de Janeiro, Brazil.
[37] Rubach S, Saur I F (1997) Onshore testing of produced water by electroflocculation. Filt. Separation, v.34, n. 8.
[38] Khemis M et al. (2005) Electrocoagulation for the treatment of oil suspensions. Trans IChemE, v. 83, n. 81, p. 50–57.
[39] Ün, Ü. T. et al.(2006) Electrocoagulation of olive mill wastewaters. Sep. Purif. Technology, v. 52, p. 136-141.
[40] Bensadok K et al.(2008) Electrocoagulation of cutting oil emulsion using aluminum plate electrodes. J. Haz. Materials, v. 152, n. 1, p.423-430.
[41] Cerqueira A A, Marques M R C, Russo C (2011) Avaliação do processo eletrolítico em corrente alternada no tratamento de água de produção. Quimica Nova, vol. 34, nº. 1, 59-63.
[42] Silva A L C (2002) Processo eletrolítico: uma alternativa para o tratamento de águas residuárias. 60f. Monografia (Especialização em Química Ambiental) - Universidade do Estado do Rio de Janeiro, Rio de Janeiro, Brazil.
[43] Kobya M et al. (2006) Treatment of potato chips manufacturing wastewater by electrocoagulation. Desalination, v. 190, n. 1-3, p. 201-211.
[44] Dimoglio A et al. (2004) Petrochemical wastewater treatment by means of clean electrochemical technologies. Clean Tech. Envir. Policy, v. 6, n. 4, p. 288-295.
[45] Merma A G (2008) Eletrocoagulação aplicada aos meios aquosos contendo óleo. 128f. Dissertação (Mestrado) - Pontifícia Universidade Católica do Rio de Janeiro, Rio de Janeiro, Brazil.
[46] Avsar Y, Kurt U, Gonullu T(2007) Comparison of classical chemical and electrochemical processes for treating rose processing wastewater. J. Haz. Materials. V. 148, n.1/2, p. 340-345.
[47] Rangel R M (2008) Modelamento da eletrocoagulação aplicada ao tratamento de águas oleosas provenientes das indústrias extrativas. 154f. Tese (Doutorado) - Pontifícia Universidade Católica do Rio de Janeiro, Rio de Janeiro, Brazil.
[48] Letterman R D, Amirtharaj A, O`Mélia C R (1999) Coagulation and flocculation. In: Water quality and treatment: a handbook of community water supplies. New York: McHill.
[49] Crespilho F N, Rezende M O O (2004) Eletroflotação: princípios e aplicações. São Carlos: Rima.

[50] Den W, Huang C (2005) Electrocoagulation for removal of silica nano-particles from chemical–mechanical-planarization wastewater. Coll. Surfaces A: Physic. Eng. Aspects, v. 254, n. 1/3, p. 81-89.

[51] Modirshahla N, Behnajady M A, Kooshaiian S (2007) Investigation of the effect of different electrode connections on the removal efficiency of Tartrazine from aqueous solutions by electrocoagulation. Dyes and Pigments, v. 74, n. 1/ 2, p. 249-257.

[52] Daneshvar N, Oladegaragoze A, Djafarzadeh N (2006) Decolorization of basic dye solution by electrocoagulation: an investigation of the effect of operational parameters. J. Haz. Materials, v. 129, p. 116-122.

[53] Daneshvar N, Sorkhabi H A, Kasiri M B (2004) Decolorization of dye solution containing Acid Red 14 by electrocoagulation with a comparative investigation of different electrode connections. J. Haz. Materials, v. 112, n. 1/2, p. 55-62.

[54] Golder A K et al. (2005) Electrocoagulation of methylene blue and eosin yellowish using mild steel electrodes. J. Haz. Materials, v. 127, n. 1/3, p.134-140.

[55] Gao P et al. (2005) Removal of chromium (VI) from wastewater by combined electrocoagulation–electroflotation without a filter. Sep. Purif Technology, v. 43, n. 2, p.117-123.

[56] Chen G (2004) Electrochemical technologies in wastewater treatment. Sep. Purif. Technology, v. 38, p.11-41.

[57] Ibrahim M Y et al. (2001) Utilization of electroflotation in remediation of oily wastewater. Sep. Science Technology, v. 36, p. 3749 – 3762.

[58] Aneshvar N et al. (2007) Decolorization of C.I. Acid Yellow 23 solution by electrocoagulation process: investigation of operational parametersand evaluation of specific energy consumption (SEEC). J. Haz. Materials, v. 148, p. 566- 572.

[59] Wimmer A C S (2007) Aplicação do processo eletrolítico no tratamento de efluentes de uma indústria petroquímica. 2007. 195f. Dissertação (Mestrado) - Pontifícia Universidade Católica do Rio de Janeiro, Rio de Janeiro, Brazil.

[60] Asselin M et al. (2008) Organics removal in oily bilgewater by electrocoagulation process. J. Haz. Materials, n. 151, p. 446–455.

[61] Fogo F C (2008) Avaliação e critérios de eficiência nos processos de tratamento de fluido de corte por eletroflotação. 103f. Dissertação (Mestrado) – Universidade de São Paulo, São Paulo, Brazil.

[62] Bande R M et al. (2008) Oil field effluent water treatment for safe disposal by electroflotation, Chem. Eng. Journal. v. 137, n. 3, p. 503-509.

[63] Ramalho A M Z (2008) Estudo de reatores eletroquímicos para remoção de Cu^{2+}, $Zn^{2=}$, Fenol e BTEX em água produzida. 86f. Dissertação (Mestrado) – Universidade Federal do Rio Grande do Norte, Natal, Brazil.

[64] Gomes E A (2007) Tratamento combinado de água produzida de petróleo por electroflocculation e processo fenton. 84f. Dissertação (Mestrado) – Universidade Tiradentes, Aracaju, Brazil.

[65] El-Naas M H et al. (2009) Assessment of electrocoagulation for the treatment of petroleum refinery wastewater. J. Envir. Management, v. 91, p.180–185.

[66] Cerqueira A A (2011) Aplicação da técnica de eletrofloculação utilizando corrente alternada de frequência variável no tratamento de água de produção da indústria do petróleo. 133f. Tese de Doutorado – Universidade do Estado do Rio de Janeiro, Rio de Janeiro, Brazil.

[67] Ferreira L H (2006) Remoção de sólidos em suspensão de efluente da indústria de papel por eletroflotação. 82f. Dissertação (Mestrado) - Universidade Estadual de Campinas, São Paulo, 2006.

[68] Irdemes S et al. (2006) The effect of current density and phosphate concentration on phosphate removal from wastewater by electrocoagulation using aluminum and iron plate electrode. Sep. Purif. Technology, v. 52, 218–223.

Chapter 2

Electrobioremediation of Hydrocarbon Contaminated Soil from Patagonia Argentina

Adrián J. Acuña, Oscar H. Pucci and Graciela N. Pucci

Additional information is available at the end of the chapter

http://dx.doi.org/10.5772/50872

1. Introduction

Bioremediation refers to the application of biological agents, typically microbes, to the removal of pollutants from an environment (e.g. through landfarming and biopiles). The effectiveness of bioremediation depends greatly on the presence of suitable microorganisms and nutrients in the subsurface. Therefore, much remains to be done in order that a generally accepted methodology can be developed for a broad range of applications [1]. One approach has been to combine bioremediation with electrokinetics (EK) into a hybrid technology, referred to as electrobioremediation (EKB). EKB uses bioremediation to degrade hydrocarbon contaminants and EK to mobilise them. EK mobilisation of the hydrocarbon products increases their bioavailability, thereby facilitating bioremediation. Whilst commonly used in the remediation of several inorganic contaminants [2,3,4,5], EK has also been successfully applied to the remediation of several soluble organic contaminants, such as phenanthrene, benzene, toluene, and phenol [6,7]. However, the efficiency of this process is severely limited when the compounds have a low solubility or bioavailability. Under these conditions, in situ flushing has the potential to improve the electrokinetically-enhanced soil-solution–hydrocarbon interaction and subsequent contaminant removal by pumping a solution directly into the subsurface of the contaminated site. Nevertheless, in situ flushing is highly dependent on the type of flushing solution employed. If the pollutant is non-ionic, it can be removed by the electroosmotic flux. However, for fine-grained soils, in which the low hydraulic permeability does not allow effective pump and treat techniques, EK remediation may be the only useful process to remove organic pollutants. Indeed, an effective hydraulic velocity of $4x10^{-7}$ m/s during EK treatment at 4 V/cm has been achieved in a soil sample with a hydraulic permeability of $5x10^{-10}$ [8]., which can be considered essentially impermeable to mechanical pumping [9].

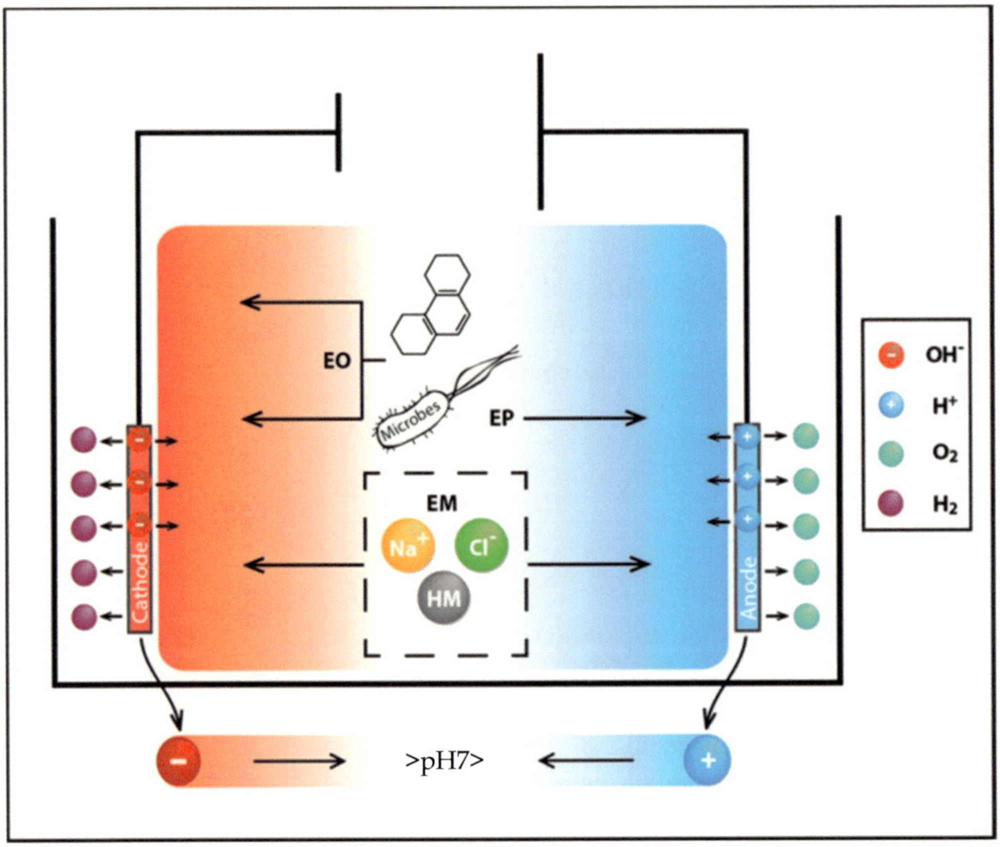

Figure 1. The effect of electrokinetic phenomena on porous soil. The application of an electric current generates hydroxide ions (OH^-) and hydrogen gas (H_2) at the cathode and hydrogen ions (H^+) and oxygen gas (O_2) at the anode. Subsequent diffusion of OH^- and H^+ introduces a pH gradient throughout the affected subsurface that in turn facilitates electrokinetic migration of soil constituents. Microbes and PAHs (exampled with phenanthrene) migrate to the cathode by electroosmosis (EO). Electronegative microbes also migrate to the anode electrophoretically (EP). Whereas electromigration (EM) dictates the migration of ions (such as sodium and chlorine) and heavy metals (HM).

The underlying mechanism of EK involves the introduction of an electric current into soil. The introduced electric current leads to the migration of contaminants via electroosmosis, electromigration, and electrophoresis. These processes occur as a consequence of the resulting pH gradient that follows the production of hydrogen ions at the anode and hydroxyl ions at the cathode (Fig.1). These phenomena cause changes in a number of soil properties [10]. Electromigration and electrophoresis result in the movement of ions, ion complexes, and charged particles, such as colloidal clay and microorganisms toward the electrode of the opposite charge. Whereas electroosmosis arises from the migration of water towards the cathode, producing an electroosmotic flow which, in turn leads to facilitates the movement of cations, hydrocarbons and microorganisms in the direction of the fluid [11]. Accordingly, changes in the available nitrogen, phosphorus and potassium in soil were observed after EK remediation [12]. The migration of electrolytes causes an increase in the same electroosmotic flux direction, whilst decreasing it at the opposite pole. The loss of moisture from EK-treated soil may also be due to warming by the passage of current, or exothermic reactions that may occur in the soil because the temperature increases between 1- 3 ºC [13]. Consequently, there must be a balance between electroosmotic migration, evaporation by heating or exothermic reactions and the supply of water at the anode.

The pH promotes interactions between metals and other compounds, that are a natural part of the soil, and regulates the availability of pollutants [2,10]. The passage of current directly into soil results in the electrolysis of water, thereby generating hydrogen ions in the anode and hydroxide ions at the cathode. This process occurs according to the following equations:

Cathode (reduction):

$$2H_2O + 2e^- \rightarrow 2OH^- + H_2$$

$$E° = -0{,}83V \text{ (alkaline)}$$

Anode (oxidation):

$$H_2O \rightarrow 2H^+ + \tfrac{1}{2} O_2 + 2e^-$$

$$E° = +1{,}23V \text{ (acid)}$$

As a result of these reactions, an acid front and a basic front are created at the anode and cathode, respectively [14]. The ideal situation occurs when the contaminant remains are dissolved in the water and not precipitated by changes in pH, when there are no changes by contact with electrodes or interactions between the contaminant and soil particles. This situation is partially fulfilled by heavy metals and some organic compounds, such as phenols or other electrically charged compounds. However, these conditions are not met by hydrocarbons present in oil as they generally have no electric charge or, if they have, it is of very low intensity. Therefore, these hydrocarbons are normally adsorbed on soil particles and are sparingly soluble in water. Under these conditions, electroosmosis is important because it allows the migration of such compounds along the path of the migrating water.

The changes induced by the application of direct current into the soil have direct effects on the microbial activity in situ. Several studies have made efforts to enhance the transport of bacteria or nutrients for effective biodegradation through the application of EKB [15;16; 17; 18, 19].

When the current-intensity is measured with different soil textures, it was found that using only large or small particles was favourable, whereas a sandy clay soil was not favourable to any of the fundamental EK processes [20].

The processes pertaining to EKB are also themselves affected by moisture, pH, chemical nature of the contaminant and zeta potential (ζ) of the soil [2]. The zeta potential is the property that determines the load of a colloid as a function of the charged surface and environment in which it is located. Fully ionisable salts are not colloids, so its ζ is very small;

the ζ in most soils is negative. With increased acidity, ζ increases such that it can reach positive values [14]. The increase in ζ impacts on the electroosmotic flow. Soil characteristics as absorbency, ion exchange buffer capacity (pH) and load surface have a marked influence on the EKB. This shows successful results in clay, fine-grained and low permeability soil. Whereas sandy soils should have an impermeable structure at a reasonable depth to allow high humidity or saturation.

The outer surface of bacterial cells possess numerous chemical groups which, at pH 7 or greater, result in an overall negative surface potential [21]. It is therefore possible to speculate about bacterial movement under the influence of an electrical field. Soil pH changes generated allow the bacteria to migrate by electrophoresis into one of the electrodes. As can be seen in Fig.1, the negatively charged membrane causes bacteria to migrate in the direction of anode [15]. Whilst the rate of migration under an electric field is quick (5 cm/h) in aqueous media, it is slow in soil, falling to ~0.8 cm /h. [17]. At low pH, the bacterial membrane charge is positive and the direction of movement is modified to the cathode. This amphoteric property is due to the complexity of surface charges on the bacterial membrane which arise from the combination of acidic (phosphates, carboxylic acids and sulphates) and basic (most notably amines) chemical groups found on the membrane surface. Consequently, it is difficult to predict their performance with biophysical parameters [21]. This difficulty is compounded as some microbes possess the ability to change their surface polarity, thus affording them some flexibility with respect to their relative migration [22]. Nonetheless, bacterial behaviour in the electric field will strongly depend on the field's intensity. The application of the electric field will result in the migration of negatively charged micro-organisms toward the anode and one-dimensional flow of pore fluid from the anode to cathode [22,23]. Importantly, a current of 40 mA at a density of 0.1-0.2 mA/cm^2 is preferable in order to achieve a one-dimensional flow of a *Pseudomonas*-loaded pore fluid from the anode to cathode [24]. However, at a current density of 0.1-0.2 mA/cm^2, the pH stabilises in the range of 2-3 at the anode destroying the acid-intolerant microbial species and in the range of 8-12 at the cathode killing the base-intolerant species. [22], thereby limiting the efficacy of the EKB process. On the other hand, the application of a recirculating buffer solution and careful regulation of electrolyte concentrations was also shown to afford some control over the pH and thereby improve distribution of a *Pseudomonas* strain which resulted in a 60% degradation of diesel over an 8 day period [24]. Maintenance of soil pH, between 5-7, is therefore necessary in order to achieve the optimum degradation of contaminants by most native soil microbes [22].

Polycyclic aromatic hydrocarbons (PAHs) are a particularly important class of pollutant as they are generated by the incomplete combustion of carbon-based fuels and are ubiquitously found in tar, oil and coal deposits [25]. Consequently, they represent one of the most widespread and abundant class of pollutants. PAHs are of particular concern as members of this class have been identified as being mutagenic, teratogenic and carcinogenic [26]. Their abundance and relative resistance to evaporation [27] makes them ideal candidates as a model contaminant through which the effectiveness of remediation technologies can be assessed.

Our main objectives were to test the effect of EKB on: (i) the removal of PAHs, (ii) to determine the increase in bioaccessibility of PAHs in soil, which would suggest improved bioremediation performance, and (iii) to evaluate the resultant change in the bacterial communities. The measured parameters included the hydrogen ion concentration (pH) values, electrical potential, bacterial count and total petroleum hydrocarbon (TPH) content and these parameters were measured along the length of each soil specimen. The results were analyzed to assess the electrokinetic remedial efficiency.

2. Material and methods

The soil used in this study was taken from a Patagonian landfill (Table 1). Samples were obtained from a depth of between 20 and 50 cm. All samples were air-dried and sieved (2 mm) prior to use in order to facilitate the even packing of the electrokinetic cells and improve the sample homogeneity.

Physical analysis		Chemical analysis (ppm)	
pH	7.4	Chloride	260
Moisture (%)	1.88	Carbonate	< 1
Apparent Density (g cm^{3-1})	1.19	Bicarbonate	101
Real Density (g cm^{3-1})	1.56	Calcium	81
Zeta potential (ζ)	-25.30	Magnesium	< 1
		Sulfate	537
Hydrocarbon analysis (%)		Nitrite	< 0.01
Total	4.51	Nitrate	24.19
Aliphatic	51	Phosphate	32.9
Aromatics	31	Iron	< 0,05
Polar	18	Ammonium	1.9

Table 1. Physics-chemical Properties of Patagonian Soil Used in This Investigation.

2.1. Electrokinetic reactor

The electrokinetic reactor used in this study was similar to that used in previous electrokinetic research [28]. Electrokinetic experiments at constant potential were carried out using an experimental apparatus for the electrokinetic tests that consisted of three main parts: soil cells, electrode compartments and power supply. A schematic of the electrokinetic reactor, and the set ups of the electrokinetic cell, is shown in Fig. 2. The electrokinetic cells consisted of a glass cell (inner dimensions: length 58cm, depth 15cm and width 15cm) that was divided into three compartments: two electrodes (10cm x 15cm x 15cm) with phosphate buffers (pH 7.8 in anode and pH of 5.8 in the cathode) using platinum electrodes inside the buffers, and a soil compartment (30cm x 15cm x 15cm). The experiments were performed using three varieties of reactor cell design: (I) in the first design, the connections between compartments were made with a NaCl agar bird channel of 1cm in diameter during one month; (II) in the second design, the electrodes were buried in the soil during one month; and (III) in the third design, the connections between compartments were done with a phosphate agar bird channel of 1cm of diameter during 150 days. All experiments were run using a constant electric field of 0.5 V/cm, and a control without electrical field was also carried out. Moisture was monitored on a weekly basis by a gravimeter method, and it was maintained at about 12%.

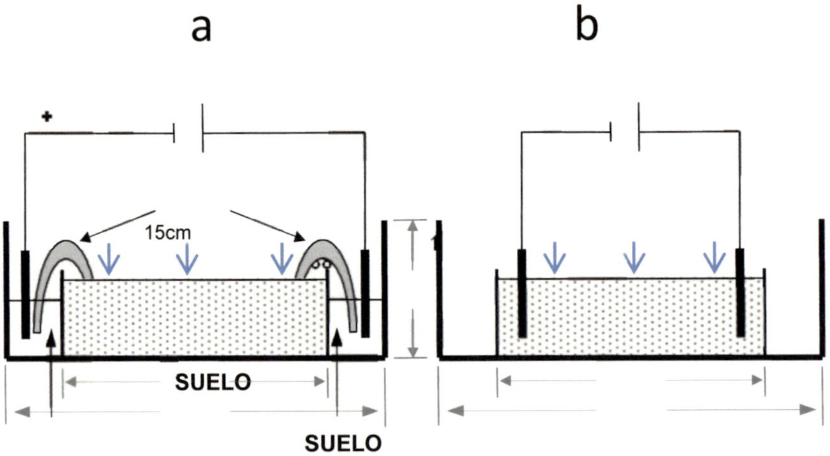

Figure 2. Design of the EK reactor used to treat polluted soil samples. Drawing A shows Experiment I, which was carried out with an agar bird channel made of Na Cl and Experiment III, which was carried out with agar bird channel made of buffer phosphate; Drawing B shows Experiment II , where the electrodes were buried in the soil.

At the end of each experiment, the soil sample was extracted from the cell and divided into 3 layers (cathode, centre and anaode), which were then divided in two samples to obtain the pH value and pollutant concentration. The pH was obtained by suspending the soil samples in de-ionised water (1:2.5, w/w) for bacterial counts, biochemical and TPH analysis.

2.2. Chemical analysis of soil samples

2.2.1. Determination of hydrocarbons via GC-analysis

Two grams of each individual sample were dissolved in 5 ml of pentane, phase separated, and percolated through 2 g of silica gel. One millilitre of the elute was carefully evaporated until dry to determine the fuel oil content of the sample. The fractions were analyzed and quantified by gas chromatography using a Varian 3800 GC, equipped with a split/splitless injector, a flame ionization detector, and a capillary column VF-5ms (30 m, 0.25 mm, 0.25 μm). The injector and detector temperatures were maintained at 200 ºC and 340 °C respectively. The Sample (1 μL) was injected in split mode and the column temperature was raised from 45 to 100 °C at a rate of 5 °C/min and a second ramp from 100 to 275 °C at a rate of 8 °C/min. The final temperature, of 275 °C, was maintained for 5 minutes.

2.2.2. *Determination of TPH content by Infrared Spectroscopy (TPH-IR)*

The soil TPH concentration was determined by infrared spectroscopy as previously described Environmental Protection Agency method [EPA 418.1]. Essentially, two grams of each individual sample were dissolved in 10 ml of carbon tetrachloride, phase separated, and percolated through 2 g of silica gel and the absorbance was measured at 2930 cm^{-1}.

2.2.3. *Determination of TPH content by Soxhlet extraction (TPH-SE).*

TPH concentration of the samples were determined by Soxhlet extractor using trichlorinethane as the extraction solvent. The extracted hydrocarbons were quantified on a mass difference basis as previously described [29] and separated into class fractions by silica gel column chromatography as formerly reported [30]. Essentially, the aliphatic, aromatic and polar oil fractions were respectively eluted using hexane (250 mL), benzene (150 mL) or 150 mL of 1:1 (v/v) chloroform-methanol.

2.3. Enumeration and isolation of aerobic bacteria.

Culturable bacteria from each sample were counted using the standard plate dilution method. One gram of soil (wet weight) was suspended in 9 ml of physiology sterile water (pH 7.2) and vortexed for 1 min at low speed. Aliquots of 100 µl of undiluted samples, and 10^{-1} to 10^{-6} dilutions were grown on TSBA (comprised of trypticase soy broth (30 g/L) and granulated agar (15 g/L)) and MBM-PGO media (comprised: NaCl (5 g/L), K_2PO_4H (0.5 g/L), $NH_4PO_4H_2$ (0.5 g/L), $(NH_4)_2SO_4$ (1 g/L), $MgSO_4$ (0.2 g/L), KNO_3 (3 g/L), $FeSO_4$ (0.05 g/L), suspended in distilled water), 30 µL of a mixture 1:1 of petroleum-diesel oil was spread on the surface once set [29] and plates incubated at 28 ºC for up to 21 days.

2.4. Chemotaxonomic analysis of soil microbe populations

The diversity of cultured sediment bacteria was determined by fatty acid methyl ester (FAME) analysis of the samples taken from the cell. FAME analysis allowed the characterization of individual bacterial colonies. Fatty acids were extracted and compared against a database, to identify isolated bacteria. From each culture plate, containing between 30 and 300 colonies, individual colonies were randomly isolated and incubated on tryptic soy broth agar for 24h. The FAMEs were extracted and analyzed by MIDI (MIDI Newark, Del., USA) as per manufacturer's instructions.

Shannon index was calculated by Sherlock (Microbial ID, version 6.0).

2.4.1. *GC parameters for MIDI analysis*

The MIDI microbial identification system (Microbial ID, Inc, Newark, NJ) was applied to separate fatty acid methyl ester using a gas chromatograph (HP 6890) equipped with a split/splitless injector, a flame ionization detector, a capillary column Ultra 2 (25 m, 0.2 mm, 0.33 µm); an automatic sampler; an integrator; and a program which identifies the fatty acids (Microbial ID 6.0 version). The injector and detector temperatures were maintained at 250 °C and 300 °C respectively. The Sample (2 μL) was injected in split mode and the column temperature was raised from 170 to 270 °C at a rate of 5 °C/min.

2.5. Statistical analysis

The mean values were compared by ANOVA test by BIOM (Applied Biostatistics Inc., NY, USA). Differences were considered significant when $P<0.05$. To identify possible similarity between FAME profiles, the data were subjected to analysis of variance using PAST [31] and Sherlock (Microbial ID, version 6.0).

3. Results

3.1. Nutrient and pH control

The use of salt bridges permitted a better regulation of pH levels in the soil, especially with the use of the phosphate buffer bridge (Fig. 3) and this did affect bacterial counts (Table 2). The introduction of phosphate in the soil benefits biodegradation due to the fact that this nutrient is necessary, especially in Patagonian soils, where the concentration of nutrients is very low. On the other hand, the NaCl bridge introduced chlorine ions into the soil, the accumulation of which results in toxic effects on the bacteria [32]. This inconvenience was observed in the electrokinetic cell and arose as a consequence of electromigration produced by the applied electric field. At the end of experiment, the concentration of K^+ and Cl^- ions following their respective migration to the cathode and anode was measured. Chlorine concentrations of 1207 mg / kg were found at the cathode whereas 836 mg / kg was observed at the anode. TheK^+ ions were found at a concentration of 50 and 42 mg / kg in the cathode and the anode, respectively. For experiment II, with electrode buried in the soil, significant changes in soil pH were observed throughout the cell (Fig. 3).

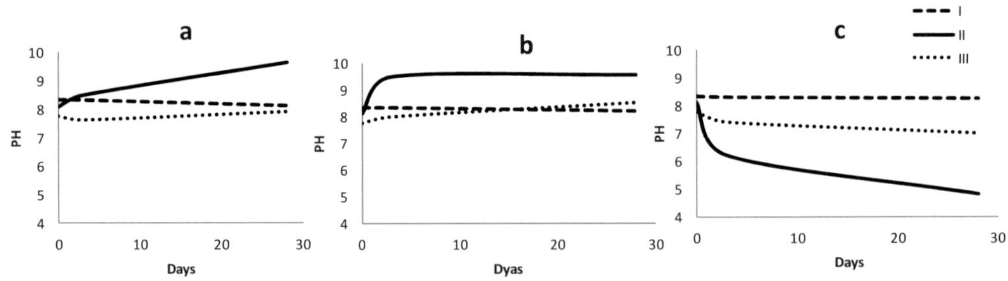

Figure 3. Effect of EKB on soil pH over time. Readings were taken from samples taken throughout the cell as follows: **a.** from the cathode, **b.** from the centre of the cell and **c.** from the anode. Data are shown for each of the reactor cell designs: I - Cells using a NaCl bridge, II – Cells in which the electrodes were placed directly into the soil and III-Cells using a PO_4^{-3} bridge.

In all three experiments, a substantial loss of moisture was observed in the electrokinetic cells. Therefore, weekly addition of water was required in order to keep the parameter between 12 and 15%. The region near the anode tended to dry, this is why the addition of water was required in all EK cells. In addition, the electrode in the soil generated a greater change in the pH and this impacted in the bacterial communities (Table 2) as well as in the moisture values.

CFUs/g				
	Initial	Anode	Centre	Cathode
Exp I	$4.12x\ 10^6$	$6.85.x\ 10^5$	2.98×10^5	2.10×10^6
Exp II	4.12×10^6	4.2×10^3	3.8×10^6	2.3×10^4
Exp III	3.10×10^7	2.77×10^7	1.62×10^8	1.04×10^8

Table 2. Obligate and Facultative Aerobic Bacteria Isolated from the EK Reactor

The results shown in Fig. 3 and Table 2 indicated that the best bridge to work with was the phosphate salt (Exp III), which could introduce nutrients to the soil and this produced an increase of bioremediation of hydrocarbons. Therefore, subsequent experiments were carried out in cells in which a phosphate bridge was placed. The nutrients introduction in Patagonian soil (Table 1), is necessary because of the soil properties [32]. In spite of this, the applied current still moves the ions. This soil needs nutrients C:N:P in the ratio 100:1:0.5 for bioremediation [32], but the anions are moved by the current, the phosphate bridges provide phosphate ions for biodegradation, and these ions accumulate in the area of the cathode.

The soil moisture contents were higher in EK remediated cells than those of the control cells due to the supply of electrolyte. However, EK remediation showed the reduced soil moisture content compared to the original soil, and the soil close to the cathode had higher moisture content than other soil, indicating the influence of electroosmosis.

Because of the electrical charge of the ions, the migration occurred and it was modified nutrient the bioavailability of phosphates. Nitrates have a relatively high mobility [33] and as shown in Table 3, the nitrates moved towards the anode. A high concentration of phosphate is seen in the cathode, probably as a result of the bridges. In accordance with previous findings [34] the values of phosphates were modified (Table 3).

	NO_3^- (ppm)	PO_4^{-3} (ppm)	NO_2^- (ppm)	NH_4^+ (ppm)
Initial	572	32.9	<0.01	1.9
Anode	1605	31.6	6.5	10.1
Cathode	479	167.7	4.1	56.7
Centre	55.9	11.4	0.96	7.6

Table 3. Electrically-Induced Migration of Nutrients in The EK Reactor.

3.2. Soil hydrocarbon content

TPH concentrations were measured by GC, TPH-IR and TPH-SE.. Analysis of the soil by TPH-IR, showed a decline in all three parts of the electrokinetics cell (Fig.4). The largest of which occurred around the anode. After 150 days, TPH-SE showed a decrease in all parts of the cells (Fig. 4). TPH analysis showed differences in values between the hydrocarbons from cells with 0.5 V/cm and the control cell, however during the first 30 days, there was there was significant difference between the values of the cathode and the anode ($P<0.05$). It is at this time that nutrients are distributed in both electrodes (Table 3). Silica gel chromatography showed changes in the percentage of the fractions of the residue (Fig. 5). The aliphatic fraction showed a decrease in all three reactor treatments. The cathode showed a good degradation. The total aromatic hydrocarbons presented a better degradation in the centre of the cell. The decrease in the percentage of aliphatic and aromatic hydrocarbons was evidenced by an increase in the relative percentage of polar hydrocarbons (Fig. 5)

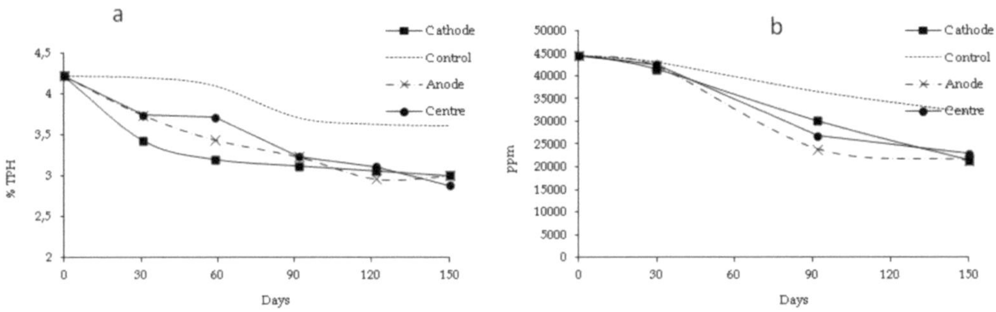

Figure 4. Effect of EKB on soil TPH levels as determined by a) TPH-SE and b)TPH-IR.

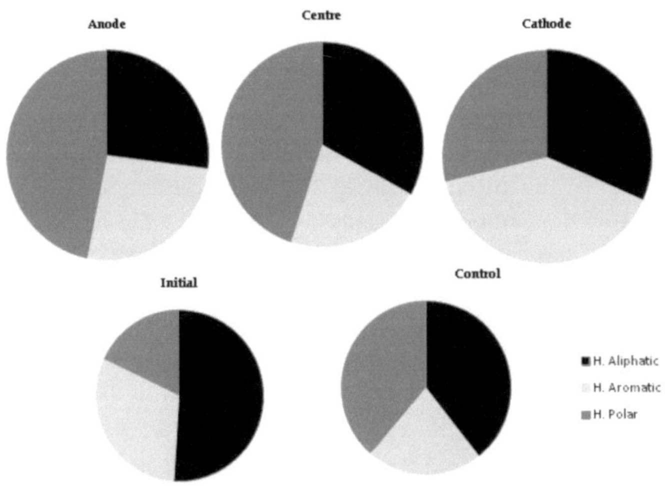

Figure 5. Effect of EKB on soil hydrocarbon content. Pie charts illustrate the relative percentages of soil hydrocarbon classes both before (initial) and, after EKB, at different locations within the cell (anode, centre and cathode).

The PAH contaminants were reduced throughout the cell, but degradation was greatest in the centre where pH was most favourable for microbial activity. Since PAHs are neutrally charged, electromigration does not work for migration of these hydrocarbons concentration profile across the soil specimens determined at the conclusion of experiments. The results show that the PAHs were degraded preferentially in anode and centre of the cell (Table 4). The phenanthrene, fluoranthene, pyrene, benzo pyrene, chrysene, benzo fluoranthene and anthracene concentrations are relatively higher at the cathode zone than at the anode zone. Considering the initial concentration of these PAHs in the soil, significant amounts ($P<0.05$) of hydrocarbon were removed by this technique.

	Initial ppm	Anode ppm	Centre ppm	Cathode ppm	Control ppm
C11	1.629	0.000	0.000	0.000	0.000
C12	6.288	0.000	0.000	0.000	0.000
C13	4.276	0.000	0.466	0.415	0.549
C14	17.254	0.779	1.008	0.868	0.947
C15	22.979	1.534	2.036	0.000	0.418
C16	42.292	3.486	3.817	3.933	4.333
C17	63.252	5.412	4.847	7.265	15.863
C18	43.783	3.955	3.068	4.823	5.882
C19	35.130	2.819	2.016	4.174	4.844
C20	70.013	1.990	1.214	3.882	4.887
C21	68.333	2.160	1.259	11.991	2.330
C22	110.245	1.555	1.726	11.931	3.408
C23	114.417	14.697	0.000	36.278	32.746
C24	84.495	14.282	2.646	20.792	117.879
C25	59.927	6.012	0.000	56.966	23.153
C26	31.689	23.686	0.000	48.624	14.803
Pristane	0.000	0.000	0.000	0.000	0.873
2-Methylnaphthalin	0.592	0.000	0.000	0.000	0.000
Methylnaphthalin	0.680	0.000	0.000	0.000	0.000
Acenaphthylen	11.653	0.000	0.000	0.000	1.615
Acenaphthen	12.896	0.000	0.000	0.871	0.991
Fluoren	26.017	0.000	0.000	0.000	0.000
Phenanthren	38.060	0.000	0.000	1.580	2.250
Anthracen	0.000	0.000	0.000	0.919	0.808
Fluoanthen	66.422	6.406	0.000	20.744	7.081
Pyren	0.000	0.000	0.000	6.100	0.000
Benzo(a)anthracen	37.186	1.182	0.000	33.196	1.234
Chrysen	1165.801	8.570	8.277	17.502	13.579
Benzo(b)fluoranthen	9.924	0.000	0.000	0.000	6.772
Benzo(k)fluoranthen	18.451	4.484	2.989	17.166	15.328
n-Alkanes	776.003	82.368	24.103	211.943	232.041
PAHs	1387.683	20.642	11.266	98.078	49.657

Table 4. Effect of EKB on Soil Content of Hydrocarbon Compounds

3.3. Bacterial counts

Maintaining the pH values suitable for the microorganisms caused the values of the bacterial counts not to experience any modification; in all cases, the drop of a logarithm was within the error of the method (Fig. 6). The bacteria did not migrate to the area of the electrodes, as stated by other authors in the case of saturated soil [35].

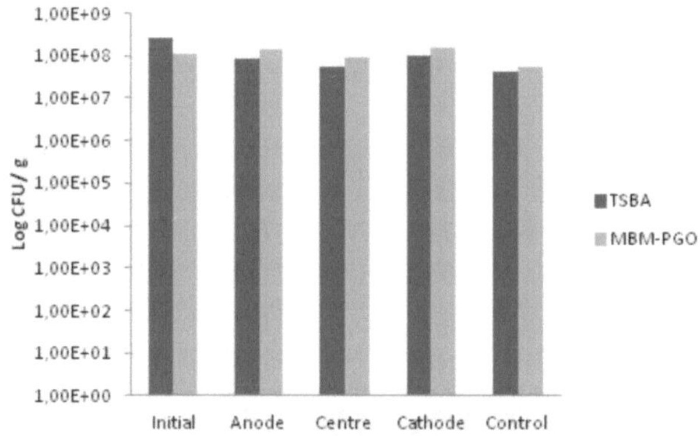

Figure 6. Bacterial count on TSBA and MBM-PGO media.

The bacterial identification was done on the bacterial count plate. In the initial sample, the genus were *Variovorax, Escherichia, Brevundimonas, Nocardia, Bordetella Mycobacterium, Rhodococcus, Acromobacter, Dierzia, Gordonia* and *Stenotrophomonas* (Fig. 7) which could grow in the agar plate in a greater number than 1 x 103. The treatment with current changed the bacterial proportion and new genus present in a small number increased their number (Fig 6).

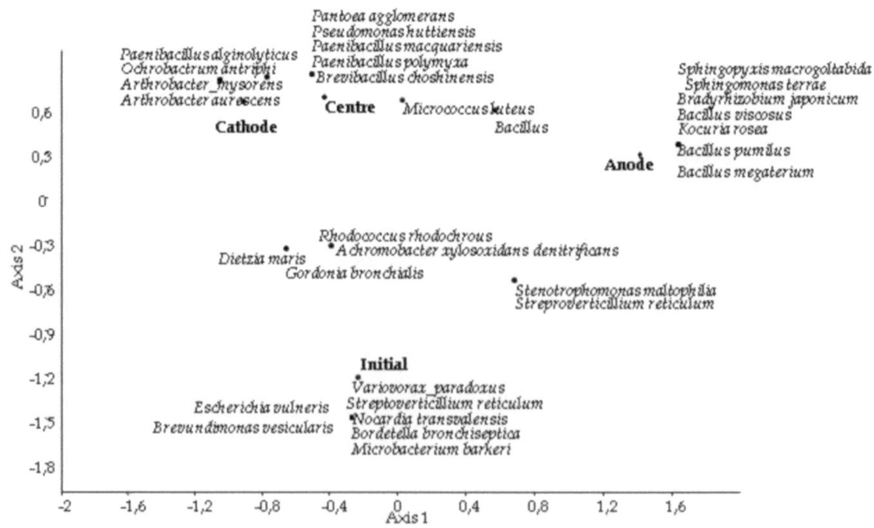

Figure 7. Statistic analysis of bacterial identification at the beginning and end of the experience.

	Initial	Cathode	Centre	Anode
Shannon_H	2.565	2.197	2.303	2.398

Table 5. Shannon indices of Soil Biodiversity

4. Discussion

The efficiency and effectiveness of electrokinetic extraction can be improved by combining the technique with other remediation technologies such as bioremediation. The results demonstrate the feasibility to degrade PAHs, a particularly problematic and toxic class of contaminant. Electrokinetics appears to be promising for a range of contaminated sites including tank bottoms which are considered dangerous waste by our legislation.

The effectiveness of the contaminant reduction during electrobioremedation depended on the bioavailability, moisture and soil pH. Among the various soil parameters, the most apparent change induced by electrobioremediation was observed in soil pH and distribution of moisture content. The bacterial number in the experiment II decreased in the anode due to the low soil pH. The soil pH is a crucial factor for microbial activity as this influences the composition and physiological characteristics of enzymes such as phosphatase, glucosidase and arylsulfatase [36]. Soil pH also affects microbial cell membrane integrity and function, and the bioavailability of nutrients and contaminants. Increased biodegradation in higher pH regions, such as cathode, has been reported for one of the PCP-degrading enzymes produced by a species of *Sphingobium* UG30 [37]. In addition, a low pH has been reported as having a major negative impact on electrobioremedation by native soil microbial communities [38,39]. Accordingly, we observed a decrease in dehydrogenase activity at locations corresponding to the acid front. In fact, in the expereince II, the soil pH changes generated by electrobioremediation were a leading cause of a decreased microbial number (Table 2). However, 0.5V/cm applied soil showed more apparent decrease in culturable bacteria compared to the soil samples of extreme initial pH [19]. This soil pH change was controlled by the use of phosphate birdge and this did not change the bacterial number between current and pH but affected the biodiversity in culturable bacteria (Fig. 7 and Table 5). The stress from the growth conditions reduces the total bacterial number and is a reason for cells entering a viable but non-culturable state [40,41]. Although electric current in EK remediation can change the bacterial membrane composition and metabolic activity, many studies revealed that weak direct current treatment has no negative effect on microbial viability and activity [42, 43, 44].

The experimental results illustrate how the application of electrokinetics to an unsaturated soil can cause major changes to the soil properties, with subsequent impact upon microbial activity and biodegradation. As expected, the lack of anodic pH control in experiment I caused progression of an acidic front through electrokinetic microcosms. However, pH control at both electrodes (experiment II) caused a large increase in moisture content in electrokinetic microcosms as the unsaturated soil absorbed water from electroosmotic flow.

In similar conditions with saturated soil this effect was not observed [18, 28, 45]. The lack of moisture change in experiment I is thought to be because the acidic pH increased the soil zeta potential, rapidly causing electroosmotic flow into the acidified region to reduce or even reverse [46]. By removing electrode fluid moisture content change was avoided but pH control had to be implemented using a regularly reversed current [47] and it may be that part of hydrocarbon fraction is eliminated in the fluids producing other contaminated residue. Changes in the concentration of pH and moisture in the soil may have mobilized pollutant fractions present in the soil in the porosity, which are often not accessible to soil microorganisms [48]. Thus, the occurrence of electroosmosis inside soil aggregates may have caused the mobilization of slowly desorbing hydrocarbons into the fast-desorbing pool, which possesses a higher bioavailability to microorganisms [49, 50]. Bioremediation in Patagonian soil with the addition of nutrients is possible; however, the remediation of PAH is problematic [29] because of their low bioavailability, which can be improved by electrobioremediation resulting in an reduction of PAH.

Maintenance of soil moisture levels proved essential, because if it decreases much it will generate problems at a metabolic level in soil bacteria. It is also necessary to apply a voltage greater than the potential in the electrodes to keep the voltage value of 0.5 $V.cm^{-1}$ thus increasing the cost of treatment energy. By working with saturated soils, the electrolyte circulation volumes, generated by electroosmosis, are contaminated is a contaminated liquid residue. One of the benefits of our system is that unsaturated soil was not observed as a leachate product of the current application. The decrease in moisture of the tanks could also be due to factors, such as the heating system or for undesirable exothermic chemical reactions [5]. Soil temperature in electrokinetics cell remained at 24 ± 3 °C throughout the 150 days of runtime. A temperature range of 24 ± 3 °C can be considered an acceptable temperature for the biodegradation of hydrocarbons by native soil microbes. These results are in accordance with results from previous studies in which an applied voltage density of 0.3 V/cm was used for this purpose [16]. This indicates that the reduction of moisture in the anode is mainly due to the presence of electroosmotic flow in the soil.

Whilst the migration of bacteria towards the anode was expected (negative surface charge being responsible for this migration), bacteria were also found to migrate towards the cathode, which may be considered surprising at first glance. This bidirectional migration of bacteria may well have resulted from the competition between two phenomena: electrophoresis and electroosmosis [51, 52]. However, other phenomena may have contributed to the migration of bacteria to the cathode. The migration of ions and water in soil toward the cathode under the influence of the electric field is one such reason. This process could create favorable conditions for bacterial growth in the area near the cathode., although the net negative charge at their surface should make them move toward the anode. Kim et al. [53], concluded "Especially the number of culturable bacteria decreased significantly and only *Bacillus* and strains in *Bacillales* were found as culturable bacteria. It is thought that the main causes of changes in microbial activities were soil pH and direct electric current". The results described here suggest that if soil parameters, electric potential difference, and electrolyte are suitably controlled based on the understanding of interaction between electrokinetics, contaminants, and indigenous microbial community, the application of electrokinetics can be a promising soil remediation technology when the contaminant is hard to degrade or its degradation is really slow.

When electric current is applied, different bacterial responses to changes in the physico-chemical properties, bioavailability, and toxic electrode-effect can be observed depending on the current, treatment period, cell type, and medium [14]. The soil pH had marked effects on microbial biomass, community structure, and response to substrate addition, and that low soil pH decreased microbial diversity and increased Gram-positive microbial communities such as *Bacillus* and *Arthrobacter* [38, 54] according with these authors the major presence of *Bacillus* were near the anode zone, and that bioelectrical reactors enhanced the metabolism of several strains in *Clostridium, Ralstonia, Pseudomonas,* and *Brevibacterium* [55].

5. Conclusion

These results indicate that biodegradation and electroosmosis can be successfully integrated to enhance PAH removal from soil, improved mobilisation of the less bioaccessible fraction of PAH with an electrokinetic pretreatment to reach lower residual levels through bioremediation. This process has the potential to provide an effective technology for the treatment of problematic soil. Whereas normal soil can be treated by 'classical' bioremedation. The optimization of these processes for a cost-effective application of the technology in situ to meet remediation will be the subject of future investigations.

Given the costs of this technology, we recommend this technique to Patagonian and other problematic soils which, after previous degradation, do not reach the levels prescribed by regulation. Optimization of this process for the removal of polyaromatic hydrocarbon from contaminated sites is the subject for future studies.

Author details

Adrián J. Acuña
Microbiology, Biochemistry Department, CEIMA,
Universidad Nacional de la Patagonia San Juan Bosco, Argentina

Oscar H. Pucci
Microbiology and Treatment of Oil Residue Extration, Biochemistry Department,CEIMA,
Universidad Nacional de la Patagonia San Juan Bosco, Argentina

Graciela N. Pucci
Treatment of Oil Residue Extration, Biochemistry Department, CEIMA,
Universidad Nacional de la Patagonia San Juan Bosco, Argentina

Acknowledgement

We acknowledge the financial support of CEIMA and UNPSJB. We sincerely thank Mirta Leiva and Miriam Robledo for their invaluable support and advice on technical aspects of this work.

6. References

[1] Chung HI, Kamon M. Ultrasonically enhanced electrokinetic remediation for removal of Pb and phenanthrene in contaminated soils. Engineering Geology. 2005;77(3-4) 233–242.

[2] Virkutyte J, Sillanpää M, Latostenmaa P. Electrokinetic soil remediation. Sci. Total Environ. 2002;289(1-3) 97-121.

[3] Ricart MT, Hansen HK, Cameselle C, Lema JM. Electrochemical treatment of a polluted sludge: different methods and conditions for manganese removal. Sep. Sci. Technol. 2004;39(15) 3679–3689.

[4] Bruell CJ, Segall BA, Walsh MT. Electroosmotic removal of gasoline hydrocarbons and TCE from clay. J. Environ. Eng. 1992;18(1) 68–83.

[5] Shapiro AP, Probstein RF. Removal of contaminants from saturated clay by electroosmosis. Environmental Science and Technology. 1993;27(2) 283–291.

[6] Saichek ER, Reddy KR. Effect of pH control at the anode for the electrokinetic removal of phenanthrene from kaolin soil. Chemosphere. 2003;51(4) 273–287.

[7] Maturi K, Reddy KR. Simultaneous removal of organic compounds and heavy metals from soils by electrokinetic remediation with a modifi ed cyclodextrin. Chemosphere. 2006;63(6) 1022–1031.

[8] Cherepy NJ, Wildenshild D. Electrolyte management for effective longterm electro-osmotic transport in low-permeability soils. Environ. Sci. Technol. 2003;37(13) 3024–3030.

[9] Polcaro AM, Vacca A, Mascia M, Palmas S. Electrokinetic removal of 2,6-dichlorophenol and diuron from kaolinite and humic acid-clay system. Journal of Hazardous Materials. 2007;148(3) 505–512.

[10] Acar YB, Alshawabkeh AN. Principles of electrokinetic remediation. Environ. Sci. Technol. 1993;27(13) 2638–2647.

[11] Bayer, E.M. & Sloyer, J.L. (1990). The electrophoretic mobility of gram-negative and gram-positive bacteria: an electrokinetic analysis. *J. Gen. Microbiol.* 1990:36(5) 867-874.

[12] Chen XJ, Shen ZM, Lei YM, Zheng SS, Ju BX, Wang WH. Effects of electrokinetics on bioavailability of soil nutrients. Soil Sci. 2006;171(8) 638-647.

[13] Shapiro AP, Probstein RF. Removal of contaminants from saturated clay by electroosmosis. Environmental Science and Technology. 1993;27(2) 283–291.

[14] Wick, L.Y.; Shi, L. & Harms, H. Electro-bioremediation of hydrophobic organic soilcontaminants:a review of fundamental interactions. *Electrochim. Acta.* 2007:52(10) 3441–3448.

[15] DeFlaun MF, Condee CW. Electrokinetic transport of bacteria. J. Hazard Mater. 1997;55(1-3) 263–277.

[16] Schmidt C, Barbosa M, Almeida M. A laboratory feasibility study on electrokinetic injection of nutrients on an organic, tropical, clayey soil. Journal Hazardous Materials. 2007;143(3) 655-661.

[17] Shi L, Muller S, Harms H, Wick LY. Effect of electrokinetic transport on the vulnerability of PAH-degrading bacteria in a model aquifer. Environ Geochem Health. 2008;30(2) 177–182.

[18] Wick LY, Mattle PA, Wattiau P, Harms H. Electrokinetic transport of PAH-degrading bacteria in model aquifers and soil. Environ Sci Technol. 2004;38(17) 4596–4602.

[19] Kim SJ, Park JI, Lee Y, Lee JIN, Yang JW. Application of a new electrolyte circulation method for the ex situ electrokinetic bioremediation of a laboratory-prepared pentadecane contaminated kaolinite. J. Hazard Mater 2005;14(118) 171–176.

[20] Mena E, Villaseñor J, Cañizares P, Rodrigo MA. Influence of soil texture on the electrokinetic transport of diesel-degrading microorganisms. Journal of Environmental Science and Health, Part A. 2011;46(8) 914-919.

[21] Bayer EM, Sloyer JL. The electrophoretic mobility of gram-negative and gram-positive bacteria: an electrokinetic analysis. J. Gen. Microbiol. 1990;136(5) 867-874.

[22] Marks RE, Acar YB, Gale RJ, Ozcu-Acar E. In-situ remediation of contaminated soils by bioelectrokinetic remediation and other competitive technologies in "Bioremediation of Contaminated Soils". Edited by Wise DL, Trantolo DJ, Cichon JE, Inyang HI, Stottmeister U, Marcel Dekker, Inc., New York-Basel. 2000; 579-606.

[23] Lou Q, Wang H, Zhang X, Qian Y. Effect of Direct Electric Current on the Cell Surface Properties of Phenol-Degrading Bacteria. Applied and environmental microbiology. 2005;71(1) 423-427.

[24] Lee HS, Lee K. Bioremediation of diesel-contaminated soil by bacterial cells transported by electrokinetics. J. Microbiol. Biotechnol. 2001;11(6) 1038–1045.

[25] Mueller JG, Cerniglia CE and Pritchard PH, Bioremediation of environments contaminated by polycyclic aromatic hydrocarbons, in *Bioremediation: Principles and Applications*, ed by Crawford RLandCrawford DL. CambridgeUniversity Press, Idaho, pp 125–194 (1996).

[26] Harvey RG, Mechanisms of carcinogenesis of polycyclic aromatic hydrocarbons. *Polycyclic Aromatic* Compounds1996(9)1–23.

[27] Warren S, Mackay D. Evaporation Rate of Spills of Hydrocarbons and Petroleum Mixtures. *Environ. Sci. Technol.* 1984 *(18)* 834-840

[28] Acuña AJ, Tonín NL, Pucci GN, Wick L, Pucci OH. Electrobioremediation of an unsaturated soil contaminated with hydrocarbon after landfarming treatment. Portugaliae Electrochimica Acta. 2010;28(4) 253-263.

[29] Pucci GN, Pucci OH. Biodegradabilidad de componentes de mezclas naturales de hidrocarburo previamente sometidas a Landfarming. Revista Argentina de Microbiología. 2003;35(2) 62-68.

[30] Acuña AJ, Pucci OH, Pucci GN. Caracterización de un proceso de biorremediación de hidrocarburos en deficiencia de nitrógeno en un suelo de la Patagonia Argentina. Ecosistema 2008;17(2) 85-93.

[31] Hammer O, Harper DAT. Paleantological Statistics version 1.34 disponible enwww.folkuio.no/ohammer/past. 2005.

[32] Pucci GN, Acuña AJ, Pucci OH. Biodegradación de hidrocarburos en la meseta patagónica, un resumen de la optimización de los parámetros a tener en cuenta. Ingeniería Sanitaria y Ambiental. 2011;115(6) 36–41

[33] Thevanayagam, S. & Rishindran, T. (1998). Injection of nutrients and TEAs in clayey soils using electrokinetics, ASCE J. *Geotech. Geoenviron. Eng.* 1998(4) 330–338.

[34] Xuejun Ch, Zhemin S, Yangming L, Shenshen Z, Bingxin J, Wenhua W. Effects of electrokinetics on bioavailability of soil nutrients. Soil science. 2006;171(8) 638-647.

[35] Suni S, Romantschuk M. Mobilisation of bacteria in soils by electro-osmosis. FEMS Microbiology Ecology. 2004;49(1) 51–57.

[36] Hinojosa MB, García-Ruiz R, Vinegla B, Carreira JA. Microbiological rates and enzyme activities as indicators of functionality in M.B. soils affected by the Aznalcollar toxic spill. Soil Biol. Biochem. 2004;36(10) 1637–1644.

[37] Habash MB, Beaudette LA, Cassidy MB, Leung KT, Hoang TA, Vogel HJ, Trevors JT, Lee H. Characterisation of tetrachlorohydroquinone reductive dehalogenase from Sphingomonas sp.UG30. Biochem. Biophys. Res. Commun. 2002;299(4) 634–640.

[38] Lear G, Harbottle MJ, Sills G, Knowles CJ, Semple KT, Thompson IP. Impact of electrokinetic remediation on microbial communities within PCP contaminated soil. Environmental Pollution. 2007;146(1) 139–146.

[39] Lear G, Harbottle MJ, van der Gast CJ, Jackman SA, Knowles CJ, Sills G., Thompson IP. The effect of electrokinetics on soil microbial communities. Soil Biology and Biochemistry. 2004;36(1) 1751–1760.

[40] Ibekwe AM, Grieve CM. Changes in developing plant microbial community structure as affected by contaminated water. FEMS Microbiol. Ecol. 2004;48(2) 239–248.

[41] Ohtomo R, Saito M. Increase in the culturable cell number of Escherichia coli during`recovery from saline stress: possible implication for resuscitation from the VBNC state. Microb. Ecol. 2001;42(2) 208–214.

[42] Lohner ST, Tiehm A. Application of electrolysis to stimulate microbial reductive PCE dechlorination and oxidative VC biodegradation. Environ. Sci. Technol. 2009;43(18) 7098-7104.

[43] Shi L, Muller S, Loffhagen N, Harms H, Wick LY. Activity and viability of polycyclic aromatic hydrocarbon-degrading Sphingomonas sp. LB126 in a DC-electrical field typical for electrobioremediation measures. Microb. Biotechnol 2008;1(1) 53–61.

[44] Tiehm A, Lohner ST, Augenstein T. Effects of direct electric current and electrode reactions on vinyl chloride degrading microorganisms. Electrochim. Acta. 2009:54(12) 3453–3459.

[45] Niqui-Arroyo JL, Bueno-Montes M, Posada-Baquero R, Ortega-Calvo JJ. Electrokinetic enhancement of phenanthrene biodegradation in creosote-polluted clay soil. Environ. Pollut. 2006;142(2) 326–332.

[46] Eykholt GR, Daniel DE. Impact of system chemistry on electroosmosis in contaminated soil. ASCE J. Geotech. Eng. 1994;120(5) 797–815.

[47] Harbottle MJ, Lear G, Sills GC, Thompson IP. Enhanced biodegradation of pentachlorophenol in unsaturated soil using reversed field electrokinetics. Journal of Environmental Management. 2009;90(5) 1893-1900.

[48] Nam K, Alexander M. Role of nanoporosity and hydrophobicity in sequestration and bioavailability: Tests with model solids. Environ. Sci. Technol. 1998;32(1) 71–74.

[49] Gómez-Lahoz C, Ortega-Calvo JJ. Effect of slow desorption on the kinetics of biodegradation of polycyclic aromatic hydrocarbons. Environ. Sci. Technol. 2005;39(22) 8776–8783.

[50] Niqui-Arroyo J, Ortaga-Calvo J. Integrating Biodegradation and Electroosmosis for the Enhanced Removal of polycyclic Aromatic Hydrocarbons from Creosote-Polluted Soils. J. Environ. Qual. 2007;36(5) 1444–1451.

[51] Liu Z, Chen W, Papadopoulos KD. Electrokinetic Movement of Escherichia coli in Capillaries. Environ. Microbiol. 1999;1(1) 99-102.

[52] Cheng-Chueh K, Papadopoulos KD. Electrokinetic Movement of Settled Spherical Particles in Fine Capillaries. Environ. Sci. Technol. 1996;30(4) 1176-1179.

[53] Kim DH, Ryu BG, Park SW, Seo CI, Baek K. Electrokinetic remediation of Zn and Ni-contaminated soil. J Hazard Mater. 2009;15(165) 501–505.

[54] Pietri JC, Brookes PC. Substrate inputs and pH as factors controlling microbial biomass, activity and community structure in an arable soil. Soil Biology and Biochemistry. 2009;41(7) 1396–1405.

[55] Thrash JC, Coates JD. Review: direct and indirect electrical stimulation of microbial metabolism. Environ. Sci. Technol. 2008;42(11) 3921–3931.

Chapter 3

Novel Formulation of Environmentally Friendly Oil Based Drilling Mud

Adesina Fadairo, Olugbenga Falode,
Churchill Ako, Abiodun Adeyemi and Anthony Ameloko

Additional information is available at the end of the chapter

http://dx.doi.org/10.5772/51236

1. Introduction

The term drilling fluids or drilling muds generally applies to fluids used to help maintain well control and remove drill cuttings (rock fragments from underground geological formations) from holes drilled in the earth. Drilling fluids are fluids used in petroleum drilling operations. These fluids are a mixture of clays, chemicals, water, oils. These fluids are used in a borehole during drilling operations for[1]:

- Hole cleaning
- Pressure control
- Cooling and lubrication of the bit
- Corrosion control (especially for oil-based muds)
- Formation damage control
- Wellbore stability maintenance
- Transmission of hydraulic energy to BHA (Bottom Hole Assembly)
- Aid in cementing operations
- Minimize environmental impact
- Inhibit gas hydrate formation in the well.
- Avoid loss of circulation and seal permeable formations.

Considering each of the uses, the primary use of drilling fluids is to conduct rock cuttings within the well. If these cuttings are not transported up the annulus between the drillstring and wellbore efficiently, the drill string will become stuck in the wellbore. The mud must be designed such that it can, carry the cuttings to surface while circulating, suspend the cuttings while not circulating, and drop the cuttings out of suspension at surface[1-5].

The hydrostatic pressure exerted by the mud column must be high enough to prevent an influx of formation fluids into the wellbore, but the pressure should not be too high, as it may fracture the formation. The instability caused by the pressure differential between the borehole and the pore pressure can be overcome by increasing the mud weight. The hydration of the clays can only be overcome by using non water-based muds, or partially addressed by treating the mud with chemicals which will reduce the ability of the water in the mud to hydrate the clays in the formation. These muds are known as inhibited muds. While drilling, the rock cutting procedure generates a lot of heat which can cause the bits, and the entire BHA (Bottom Hole Assembly) wear out and fail, and the drilling muds help in cooling and lubricating the BHA. These fluids also help in powering the bottom hole tools. In cementing operations, drilling fluids are used to push and pump the cement slurry down the casing and up the annular space around the casing string in the hole.

The drilling fluid must be selected and or designed so that the physical and chemical properties of the fluid allow these functions to be fulfilled. However, when selecting the fluid, consideration must also be given to[5-6]:

- The environmental impact of using the fluid
- The cost of the fluid
- The impact of the fluid on production from the reservoir

2. Classification of drilling fluids

Drilling fluids are classified according to the continuous phase[1,3]

- The WBM (Water Based Muds), with water as the continuous phase.
- The OBM (Oil Based Muds), with oil as their continuous phase.
- The Pneumatic fluids (with gases or gas-liquid mixtures as their continuous phase)

This chapter narrows our focus to oil based drilling fluids (OBM).

In general, OBM are drilling fluids which have oil as their dominant or continuous phase. A typical OBM has the following composition:

Clays and sand about 3%, Salt about 4%, Barite 9%, Water 30%, Oil 50-80%.

OBM have a whole lot of advantages over the conventional WBM. This is due to the various desirable rheological properties that oils exhibit. Since the 1930s, it has been recognized that better productivity is achieved by using oil rather than water as the drilling fluid. Since the oil is native to the formation it will not damage the pay zone by filtration to the same extent as would a foreign fluid such as water. We shall outline some of the desirable properties of oil based muds, which include[4]:

1. **Shale Stability: OBM** are most suited for drilling shaly formations. Since oil is the continuous phase & water is dispersed in it, this case results in non-reactive interactions with shale beds.
2. **Penetration Rates: OBM** usually allow for increased penetration rates.

3. **Temperature:** OBM can be used to drill formations where BHT (Bottom Hole Temperatures) exceed water based mud tolerances. Sometimes up to over 1000 degrees rankine.
4. **Lubricity:** OBM produce thin mud cakes, and the friction between the pipe and the well bore is minimized, thus reducing the pipe differential sticking. Especially suitable for highly deviated and horizontal wells.
5. Ability to drill low pore pressured formations is accomplished, since the mud weight can be maintained at a weight less than that of water (as low as 7.5 ppg).
6. **Corrosion control:** Corrosion of pipes is reduced since oil, being the external phase coats the pipe. This is due to the fact that oils are non conductive, thermally stable, and more often, do not permit microbial growth.
7. OBM can be re used, and can also be stored for a long period of time since microbial activity is suppressed.

The basic kind of oil used in formulating OBM is the diesel oil, which has been in existence for a long time, but over the years, diesel oil based muds have posed various environmental problems.

Water-based muds (WBMs) are usually the mud of choice in most drilling operation carried out in sandstone reservoir, however some unconventional drilling situations such as deeper wells, high temperature/pressure formation, deepwater reservoir, alternative shale-sand reservoir and shale resource reservoir require use of other mud systems such as oil based mud to provide acceptable drilling performance[5-8].

OBM is needed where WBM cannot be used especially in hot environment and salt beds where formation compositions can be dissolved in WBM. OBM have oil as their base and therefore more expensive and require more stringent pollution control measures than WBM.

It is imperative to propagate the use of environmentally friendly and biodegradable sources of oil to formulate our OBM, thereby making it less expensive and environmentally safe and equally carry out the basic functions of the drilling mud such as maintenance of hydrostatic pressure, removal of cuttings, cooling and lubricating the drill string and also to keep newly drilled borehole open until cementing is carried out.

2.1. Background

Environmental problems associated with complex drilling fluids in general, and oil-based mud (OBM) in particular, are among the major concerns of world communities. Among others are the problems faced by some host communities in the Niger Delta region of Nigeria. For this reason, the Environmental Protection Agency (EPA) and other regulatory bodies are imposing increasingly stringent regulations to ensure the use of environmentally friendly muds[7-8].

Throughout the 1970s and 1980s, the EPA and other regulatory bodies imposed environmental laws and regulations affecting all aspects of petroleum-related operations from exploration, production and refining to distribution. In particular, there has been increasing pressure on oil and gas industry stakeholders to find environmentally acceptable alternatives to OBMs. This has been reflected in the introduction of new legislation by government agencies in almost every part of the world.

The researches and surveys conducted came up with possibilities of having environmentally friendly oil based mud. Stakeholders in the oil and gas industry have been tasked with the challenge of finding a solution to this problem by formulating optimum drilling fluids and also reduce the handling costs and negative environmental effects of the conventional diesel oil based drilling fluid. An optimum drilling fluid is one which removes the rock cuttings from the bottom of the borehole and carries them to the surface, hold cuttings and weight materials in suspension when circulation is stopped (e.g during shut in), and also maintain pressure. An optimum drilling fluid also does this at minimum handling costs, bearing in mind the HSE (Health, Safety, Environment) policy in mind[6].

In response to the harmful effects of diesel oil on the environment and on the ozone layer (as a result of the emission of greenhouse gases), researches and surveys have gone on in the past two to three decades, and have come up with mud formulations based on the use of plant oils as diesel substitutes. Over the years, plant oils have become increasingly popular in the raw materials market for diesel substitutes. The most popular being: Rapeseed oil, Jatropha oil, Mahua oil, Cottonseed oil, Sesame oil, Soya bean oil, palm oil etc. This brings about the importance of agro allied intervention in the energy industry. Hence, the contribution of non-edible oils such as jatropha oil, canola oil, algae oil, moringa seed oil and Soapnut will be significant as a plant oil source for diesel substitute production.

This chapter describes the formulation of environmental friendly oil based mud (using plant oil such as jatropha oil, algae oil and moringa seed oil) that can carry out the same functions as diesel oil based drilling fluid and equally meet up with the HSE (Health, Safety and Environment) standards. Mud tests have been carried out at standard conditions on each plant oil sample so as to ascertain the rheological properties of the drilling fluid formulations. The conventional diesel oil based mud would serve as control.

2.2. Motivation

Drilling mud is in varying degrees of toxicity. It is difficult and expensive to dispose it in an environmentally friendly manner. Protection of the environment from pollutants has become a serious task. In most countries like Nigeria, the drilling fluids industries have had numerous restrictions placed on some materials they use and the methods of their disposal. Now, at the beginning of the 1990's, the restrictions are becoming more stringent and restraints are becoming worldwide issues. Products that have been particularly affected by restrictions are oil and oil-based mud. These fluids have been the mud of choice for many environments because of their better qualities. Initially, the toxicity of oil-based fluids was reduced by the replacement of diesel oil with low-aromatic mineral oils. In most countries today, oil-based mud may be used but not discharged in offshore or inland waters. Potential liability, latent cost, and negative publicity associated with an oil-mud spill are economic concerns. Consequently, there is an urgent need for the drilling fluids industry to provide alternatives to oil-based mud.

2.3. Methodology of the study

Four different mud samples were mixed, and the base fluid was varied. The base fluids were algae, moringa, diesel and jathropha oils used in formulating the muds in an oil water ratio of 70:30, where diesel based mud served as the control.

The following equipment and materials were used to carry out the experiment:

Materials	Equipment
1. Pulverized bentonite	1. Weighing balance
2. Barite	2. Retort
3. Diesel oil	3. Halminton Beach Mixer
4. Canola oil	4. Condenser
5. Castor oil	5. Mud balance
6. Jatropha seeds	6. Round bottom flask
7. Water	7. Rotary viscometer
8. n-hexane	8. Resistivity meter
9. Filter paper	9. API filter press
10. Threads	10. pH meter
11. Universal pH paper strips	11. Soxhlet extractor
12. Algae	12. Heating mantle
	13. Vernier Caliper
	14. Reagent bottles

Table 1. Materials and Apparatus required

2.4. Experimental procedure

The plant seeds (jatropha, moringa and algae) were collected from the western part of Nigeria, peeled and dried in an oven at about 55°C for seventy minutes. The dried seeds were then de-hulled, to remove the kernels. The brownish inner parts of the kernels were ground in a blender (to increase the surface area for the reaction).

2.5. Extraction

The method employed in this study is solvent extraction. Solvent extraction is a process which involves extracting oil from oil-bearing materials by treating it with a low boiling point solvent as opposed to extracting the oils by mechanical pressing methods (such as expellers, hydraulic presses, etc.). The solvent extraction method recovers almost all the oils and leaves behind only 0.5% to 0.7% residual oil in the raw material. Here the equipment used was the Soxhlet extractor. A Soxhlet extractor is a piece of laboratory apparatus invented in 1879 by Franz von Soxhlet. It was originally designed for the extraction of a lipid from a solid material.

Figure 1. Soxhlet extractor assembly.

The extraction procedure is given below:

1. 50g of crushed plant seeds were measured out, and tied in filter papers.
2. The sample was loaded into the main chamber of the Soxhlet extractor and poured in about 300ml of n-Hexane through the main chamber.
3. The chamber is fitted into a flask containing 300ml of n-Hexane.
4. The heating mantle was turned on and the system was left to heat at 70° C. The solvent was heated to reflux. The solvent vapour travelled up a distillation arm, and flooded into the chamber housing the solid wrapped in filter papers. The condenser condensed the solvent vapour, and the vapour dripped back down into the chamber housing the solid material.
5. Then at a certain level, the siphon emptied the liquid into the flask.
6. This cycle was repeated until the sample in the chamber changed colour to a considerable extent, and collected the fluid mixture in glass reagent bottles.
7. The mixture was separated via the use of simple distillation, as shown in the set up in Fig. 2.
8. The distillation took place at 70°C; the hexane was recovered and re-used while the oil was stored.

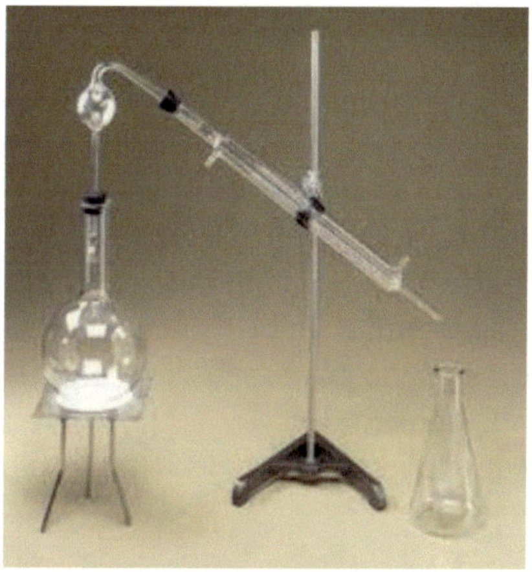

Figure 2. Set-up for distillation.

3. Mud preparation

The densities of the various base fluids (water, algae oil, moringa oil, jatropha oil and diesel) were measured using the mud balance shown in diagram 3

1. Using the weighing balance, the various quantities of materials as shown in Table 2 below were measured.
2. The quantities of water and oil were measured using measuring beakers.
3. Using the Hamilton Beach Mixer, the measured materials were thoroughly mixed until a homogenous mixture was obtained.
4. The mud samples were aged for 24 hours.

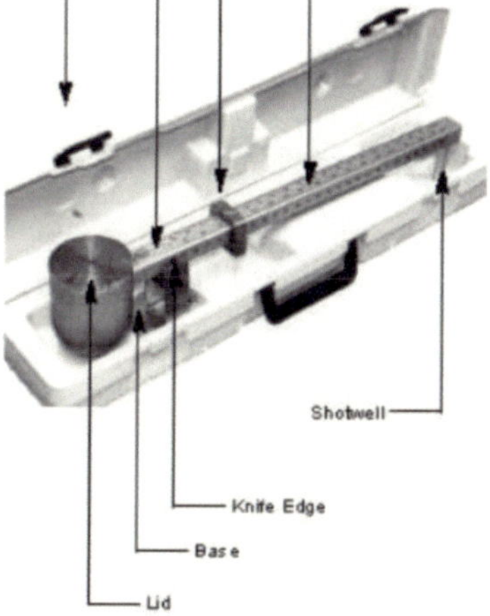

Figure 3. Mud Balance

3.1. Density

1. The aged mud samples were agitated for 2 minutes using the Hamilton Beach Mixer.
2. The clean, dry mud balance cup was filled to the top with the newly agitated mud.
3. The lid was placed on the cup and the balance was washed and wiped clean of overflowing mud while covering the hole in the lid.
4. The balance was placed on a knife edge and the rider moved along the arm until the cup and arm were balanced as indicated by the bubble.
5. The mud weight was read at the edge of the rider towards the mud cup as indicated by the arrow on the rider and was recorded.
6. Steps 2 to 5 were repeated for the other samples.

3.2. Viscosity

7. The mud was poured into the mud cup of the rotary viscometer shown in Diagram 4, and the rotor sleeve was immersed exactly to the fill line on the sleeve by raising the platform. The lock knot on the platform was tightened.
8. The power switch located on the back panel of the viscometer was turned on.
9. The speed selector knob was first rotated to the stir setting, to stir the mud for a few seconds, and it was rotated at 600RPM, waiting for the dial to reach a steady reading, the 600 RPM reading was recorded.
10. The above process was repeated for 300 RPM, 200 RPM, 100 RPM, 60 RPM, 30 RPM and 6 RPM.
11. Steps 7 to 10 were repeated for other samples.

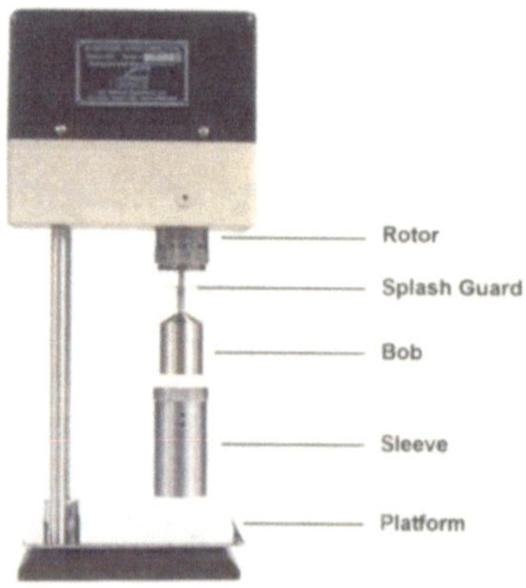

Figure 4. Rotational Viscometer

3.3. Gel strength

12. The speed selector knob was then rotated to to stir the mud sample for a few seconds, then it was rotated to gel setting and the power was immediately shut off.

13. As soon as the sleeve stopped rotating, the power was turned on after 10 seconds and 10 minutes respectively. The maximum dial was recorded for each case.
14. Steps 12 and 13 were repeated for other samples.

3.4. Mud filtration properties

15. The assembly is as shown in fig 5
16. Each part of the cell was cleaned, dried and the rubber gaskets were checked.
17. The cell was assembled as follows: base cap, rubber gasket, screen, filter paper, rubber gasket, and cell body.

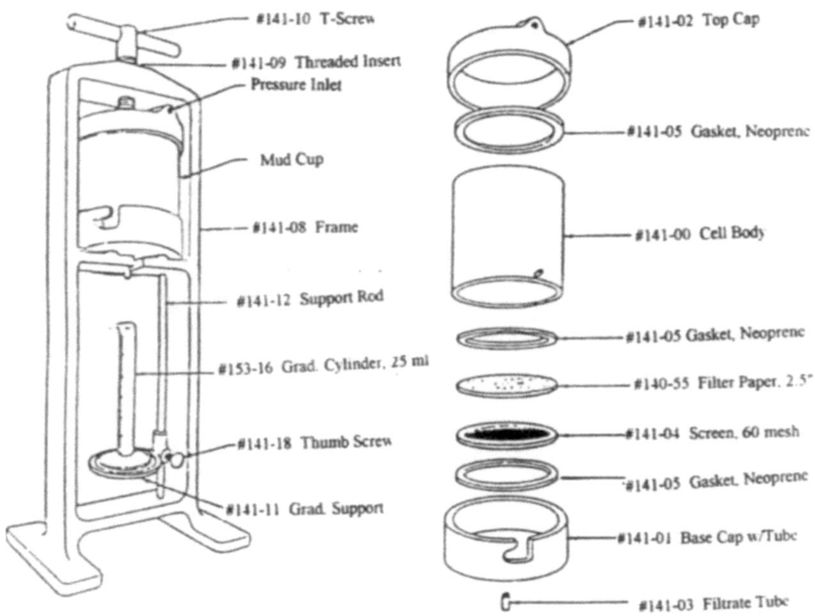

Figure 5. API Filter Press

18. A freshly stirred sample of mud was poured into the cell to within 0.5 inch (13 millimeters) to the top in order to minimize contamination of the filtrate. The top cap was checked to ensure that the rubber gasket was in place and seated all the way around and complete the assembly. The cell assembly was placed into the frame and secured with the T-screw.
19. A clean dry graduated glass cylinder was placed under the filtrate exit tube.
20. The regulator T-screw was turned counter-clockwise until the screw was in the right position and the diaphragm pressure was relieved. The safety bleeder valve on the regulator was put in the closed position.
21. The air hose was connected to the designated pressure source. The valve on the pressure source was opened to initiate pressurization into the air hose. The regulator was adjusted by turning the T-screw clockwise so that a pressure was applied to the cell in 30 seconds or less. The test period begins at the time of initial pressurization.

22. At the end of 30 minutes the volume of filtrate collected was measured. The air flow through the pressure regulator was shut off by turning the T-screw in a counter-clockwise direction. The valve on the pressure source was then closed and the relief valve was carefully opened.
23. The assembly was then dismantled, and the mud was removed from the cup.
24. The filter cake was measured using a vernier caliper, and the measurements were recorded.
25. The above procedures were carried out for the other mud samples.

3.5. Hydrogen ion concentration (pH)- Colorimetric paper method

26. A short strip of pH paper was placed on the surface of the sample.
27. After the color of the test paper stabilized, the color of the upper side of the paper, which had not contacted the mud, was matched against the standard color chart on the side of the dispenser.
28. Steps 26 and 27 were carried out on other samples.

4. Toxicity test

29. After the oil based mud samples have been formulated, each is then tested on a growing plant (that is on beans seedling), to see the effects on the plant growth and the living organisms in the soil. Bean seed was planted and exposed to 100ml of three different mud samples, with the following base fluids; diesel, canola and jatropha, the growth rate was measured, and the number of days of survival.

4.1. Results of density measurements

The results as obtained from measurements of density using the mud balance are contained in Table 2 below.

SAMPLE	MEASURED DENSITY (ppg)	CALCULATED DENSITY (ppg)	ERROR	Barite (g)
Diesel	8.26	8.261	0.01	119.1
Algae	7.81	7.815	0.005	126.5
Jatropha	8.32	8.326	0.06	154.5
Moringa	8.30	8.307	0.007	149.3
Canola	8.47	8.470	0	150.6

Table 2. Mud density values

Mud density ϱ is calculated using eqn $\rho_m = \frac{M_{Ben} + M_{Oil} + M_{Water}}{V_{Ben} + V_{Oil} + V_{Water}}$

e.g for Jatropha

$$\rho_{m,J} = \frac{0.110231 + 0.38040768 + 0.76742464}{0.0924608 + 0.0528344 + 0.005079769585} = 8.326 \text{ ppg}$$

From the above table, the error differences between the calculated and measured densities all lie below 0.1, thus the readings obtained using the mud balance have a high accuracy. It also showed that the denser the base oil, the higher the amount of barite needed to build.

4.2. Viscosity and gel strength results

Viscosity readings obtained from the experiment carried out on the rotary viscometer are contained in Table 3.

The dial reading values (in lb/100ft^2) are tabulated against the viscometer speeds in RPM.

Viscosity values are calculated with equations

Apparent viscosity= Dial Reading at 600RPM (θ_{600})/2

Dial speed (RPM)	Diesel	Algae	Jatropha	Moringa	Canola
600	185	122	154	169	128
300	170	114	133	158	120
200	169	96	124	149	115
100	163	88	114	143	114
60	152	82	107	140	113
30	143	74	98	136	111
6	122	62	92	120	110
3	81	55	76	79	60

Table 3. Viscometer Readings for Diesel, Jatropha and Canola OBM's

Rheological Properties	Diesel	Algae	Jatropha	Moringa	Canola
Plastic Viscosity	15	8	21	11	8
Apparent Viscosity	92.5	61	77	84.5	64
Gel Strength	50/51	52/43	54/55	52/53	60/72

Table 4. Plastic Viscosities, Apparent Viscosities, Gel Strength,

Diesel OBM had the highest apparent viscosity, followed by Moringa, then Jatropha, Canola and algae OBM's

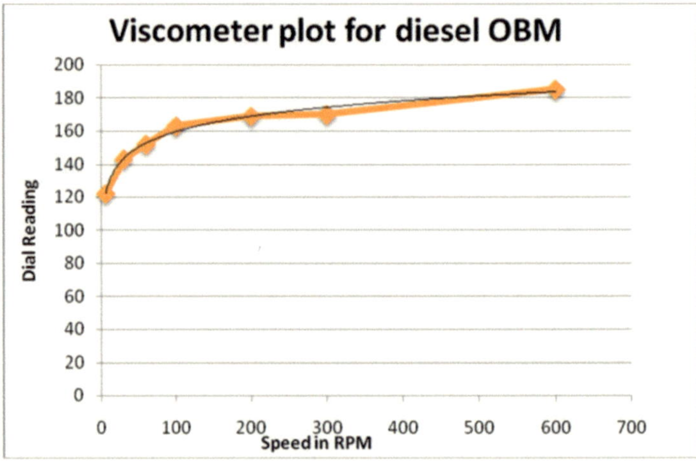

Figure 6. Viscometer Plot for Diesel OBM

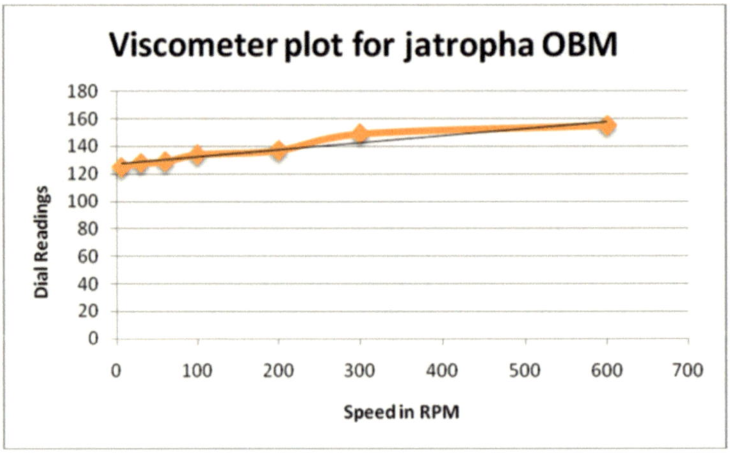

Figure 7. Viscometer Plot for Jatropha OBM

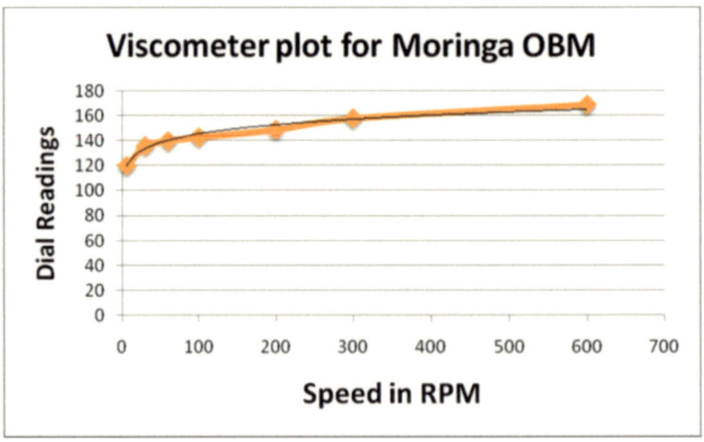

Figure 8. Viscometer Plot for Moringa OBM

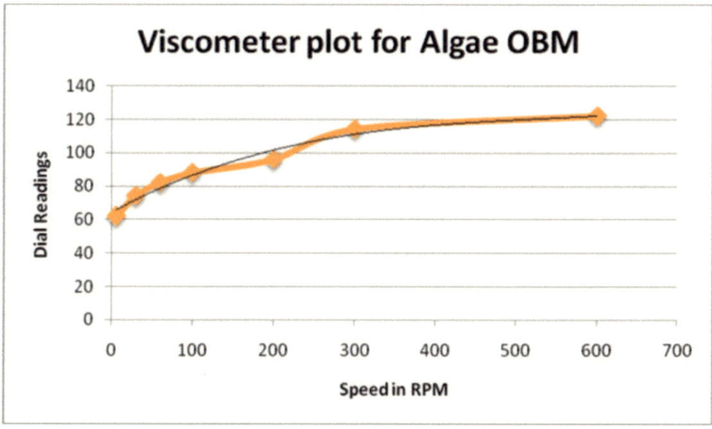

Figure 9. Viscometer Plot for algae OBM

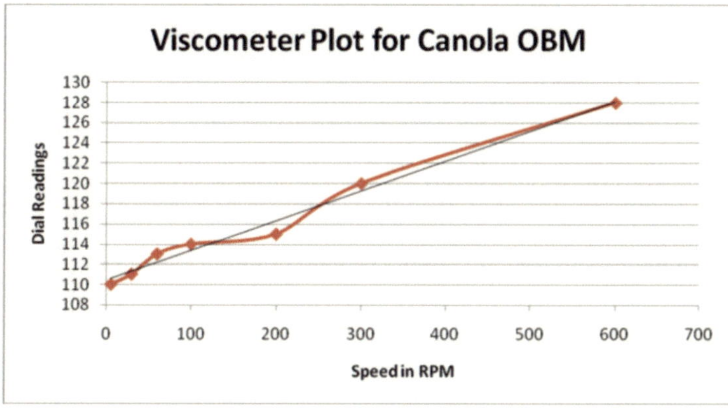

Figure 10. Viscometer Plot for Canola OBM

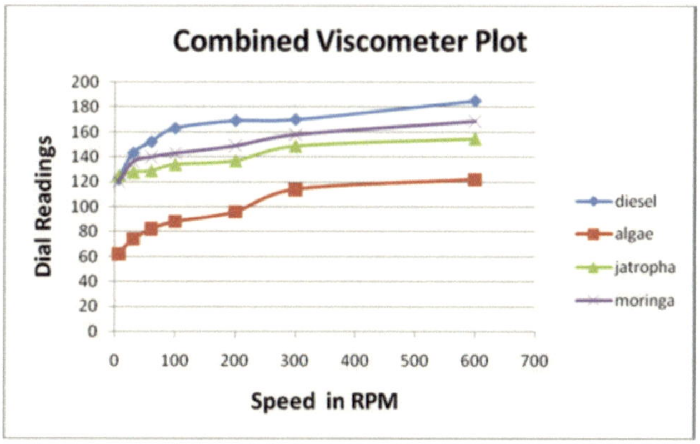

Figure 11. Combined viscometer plot for Diesel, Algae, and jatropha OBM's

It can be seen that the plots on Figures 6 to 11, generated from the dial readings of all the mud samples are similar to the Bingham plastic model. This goes to prove that the muds have similar rheological behaviour.

However, not all the lines of the plot are as straight as the Bingham plastic model. This can be explained by a number of factors such as: possible presence of contaminants, and the possibility of behaving like a different model such as Herschel Bulkley.

A Bingham plastic fluid will not flow until the shear stress τ exceeds a certain minimum value τ_y known as the yield point[9] (Bourgoyne et al 1991). After the yield has been exceeded, the changes in shear stress are proportional to changes in shear rate and the constant of proportionality is known as the plastic viscosity μ_p.

From Figures, the yield points of the different muds can be read off. The respective yield points are the intercepts on the vertical (shear stress) axes.

For reduced friction during drilling, algae OBM gives the best results, followed by Jatropha OBM then moringa OBM.

This means Diesel OBM offers the greatest resistance to fluid flow. Algae, Jatropha, Moringa and Canola OBM's pose better prospects in the sense that their lower viscosities will mean less resistance to fluid flow. This will in turn lead to reduced wear in the drill string[10].

4.3. Mud filtration results

The filtration tests were carried out at 350 kPa due to the low level of the gas in the cylinder.

The mud cakes obtained from the API filter press exhibited a slick, soft texture.

From Table 5 and Figures 12 to 15, we can infer that Diesel OBM had the highest rate of filtration and spurt loss. Comparing this to a drilling scenario, this means that the mud cake from Diesel OBM is the most porous, and the thickest.

From these inferences, we can see that Algae, Jatropha, Moringa and Canola OBM's are better in filtration properties than Diesel OBM as inferred from thickness and filtration volumes.

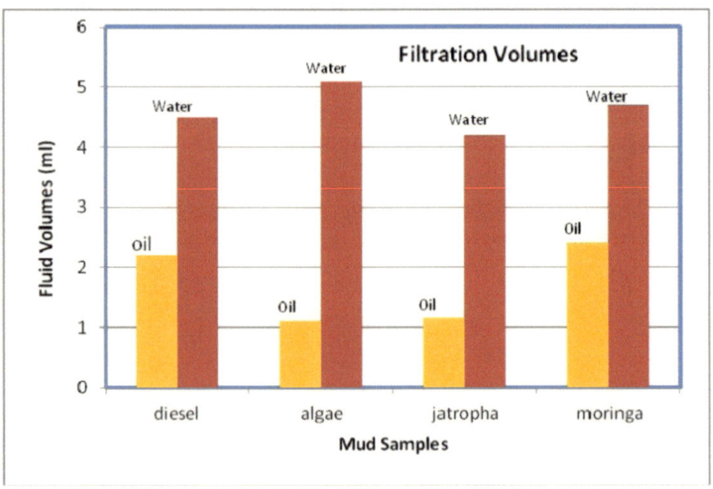

Figure 12. Filtration Volumes for Diesel, Algae, Jatropha and Moringa OBM's

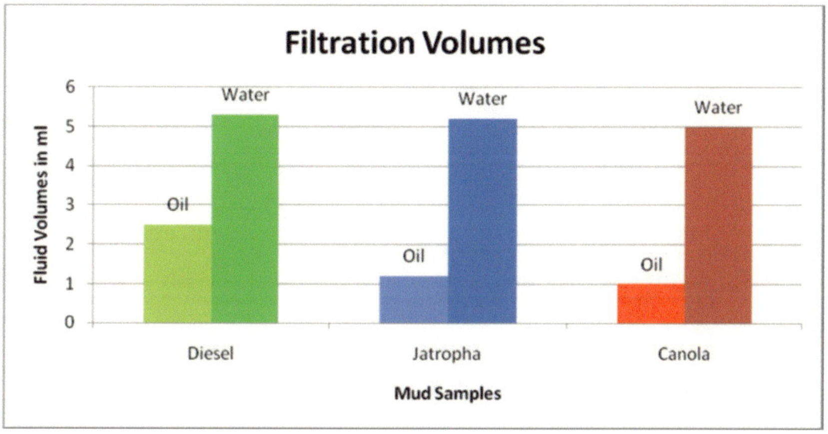

Figure 13. Filtration Volumes for Diesel, Jatropha and Canola OBM's

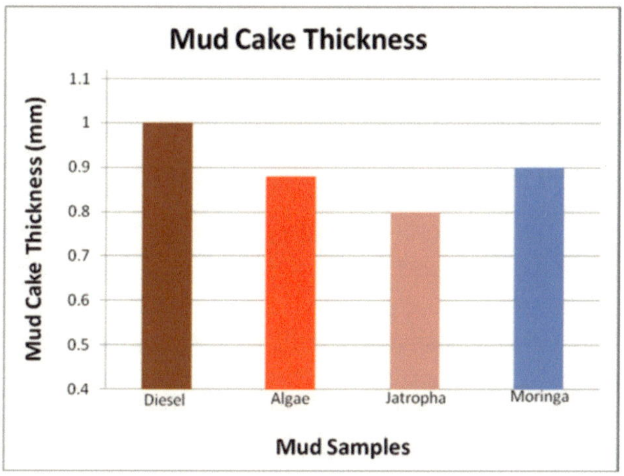

Figure 14. Mud Cake Thicknesses for Diesel, Algae, Canola OBM's

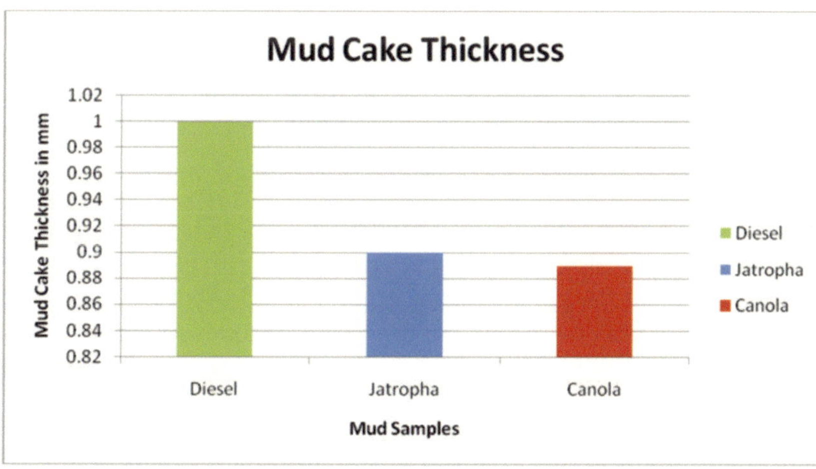

Figure 15. Mud Cake Thicknesses for Diesel, Jatropha and Canola OBM's

Filtration Properties	DIESEL	ALGAE	JATROPHA	MORINGA	Canola
Total Fluid Volume	6.9ml	6.2ml	6.3ml	7.2ml	6.0 ml
Oil volume	2.3ml	1.1ml	1.1ml	2.5ml	1.0 ml
Water Volume	4.6ml	5.1ml	4.2ml	4.7ml	4.3 ml
Cake Thickness	1.0mm	0.9mm	0.8mm	0.9mm	0.78mm

Table 5. Mud Filtration Results

Problems caused as a result of excessive thickness include[4]:

i. Tight spots in the hole that cause excessive drag.
ii. Increased surges and swabbing due to reduced annular clearance.
iii. Differential sticking of the drillstring due to increased contact area and rapid development of sticking forces caused by higher filtration rate.
iv. Primary cementing difficulties due to inadequate displacement of filter cake.
v. Increased difficulty in running casing.

The problems as a result of excessive filtration volumes include[4]:

i. Formation damage due to filtrate and solids invasion. Damaged zone too deep to be remedied by perforation or acidization. Damage may be precipitation of insoluble compounds, changes in wettability, and changes in relative permeability to oil or gas, formation plugging with fines or solids, and swelling of in-situ clays.
ii. Invalid formation-fluid sampling test. Formation-fluid flow tests may give results for the filtrate rather than for the reservoir fluids.
iii. Formation-evaluation difficulties caused by excessive filtrate invasion, poor transmission of electrical properties through thick cakes, and potential mechanical problems running and retrieving logging tools.
iv. Erroneous properties measured by logging tools (measuring filtrate altered properties rather than reservoir fluid properties).
v. Oil and gas zones may be overlooked because the filtrate is flushing hydrocarbons away from the wellbore, making detection more difficult.

4.4. Hydrogen ion potential results

Drilling muds are always treated to be alkaline (i.e., a pH > 7). The pH will affect viscosity, bentonite is least affected if the pH is in the range of 7 to 9.5. Above this, the viscosity will increase and may give viscosities that are out of proportion for good drilling properties. For minimizing shale problems, a pH of 8.5 to 9.5 appears to give the best hole stability and control over mud properties. A high pH (10+) appears to cause shale problems.

The corrosion of metal is increased if it comes into contact with an acidic fluid. From this point of view, the higher pH would be desirable to protect pipe and casing (Baker Hughes, 1995).

The pH values of all the samples meet a few of the requirements stated but Diesel OBM with a pH of less than 8.5 does not meet with specification. Algae, Jatropha, Moringa and Canola OBM's show better results since their pH values fall within this range.

Type of Oil	DIESEL	ALGAE	JATROPHA	MORINGA
pH Value	8	9	8.5	9

Table 6. pH Values

4.5. Results of cuttings carrying index

Only three drilling-fluid parameters are controllable to enhance moving drilled solids from the wellbore:Apparent Viscosity (AV) density (mud weight [MW]), and viscosity. Cuttings Carrying Index (CCI) is a measure of a drilling fluid's ability to conduct drilled cuttings in the hole. Higher CCI's, mean better hole cleaning capacities.

From the Table, we can see that Jatropha OBM showed best results for CCI iterations.

	Diesel	Jatropha	Canola
CCI	15.901	19.067	17.846

Table 7. Cuttings Carrying Indices (CCI's)

4.6. Pressure loss modeling results

The Bingham plastic model is the standard viscosity model used throughout the industry, and it can be made to fit high shear- rate viscosity data reasonably well, and is generally associated with the viscosity of the base fluid and the number, size, and shape of solids in the slurry, while yield stress is associated with the tendency of components to build a shear-resistant.

	Diesel	Jatropha	Canola
Drill Pipe	829	277.39	250.65
Drill Collar	177.35	173.75	157.0
Drill Collar (Open)	161.35	158.15	142.9
Drill Pipe (Open)	14.1	13.81	12.48
Drill Pipe (Cased)	9.28	9.10	8.22
Total	1191.98	706.45	571.25

Table 8. Bingham Plastic Pressure Losses in Psi

It can be seen from the table that Jatropha and Canola OBM's gave better pressure loss results than Diesel OBM as a result of lower plastic viscosities, and hence should be encouraged for use during drilling activities.

4.7. Result of the toxicity measurements

Samples of 100ml of each of the selected oils were exposed to both corn seeds and bean seed and the no of days which the crop survived are as indicated in Figure 16. The growth rate was also measured i.e the new length of the plant was measured at regular time intervals. For the graph of toxicity of diesel based mud the reduced growth rate indicates when the leaves began to yellow, and the zero static values indicate when the plant died.

From the results indicated by the figure 16, it can be concluded that jatropha oil has less harmful effect on plant growth compared to canola and diesel. Bean seeds were planted and after one week, they were both exposed to 100ml of both jathropha formulated mud and diesel formulated mud. The seeds exposed to jatropha survived for 18 days, while that exposed to diesel mud survived for 6 days and then withered. When the soil was checked, there was no sign of any living organisms in diesel mud sample while that of the jatropha mud, there were signs of some living organisms such as earth worms, and other little insects. This shows that jatropha mud sample is environmentally safer for both plants and micro animals than diesel mud sample.

From the figure 17, it can be seen that the seeds exposed to jatropha had the highest number of days of survival which indicates its lower toxicity while that of diesel had the lowest days of survival which indicates its high toxicity. The toxicity of diesel can be traced to high aromatic hydrocarbon content. Therefore, replacements for diesel should either eliminate or minimize the aromatic contents thereby making the material non toxic or less toxic. Biodegradation and bioaccumulation however depend on the chemistry of the molecular character of the base fluids used. In general, green material i.e plant materials containing oxygen within their structure degrade easier.

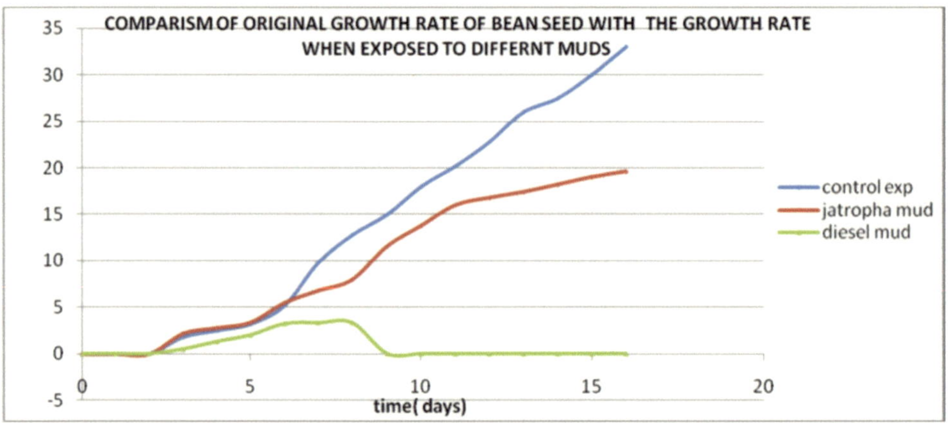

Figure 16. Comparison of Growth Rate Curve of Different Mud Types

4.8. Results of density variation with temperature

Densities were measured for the various samples at temperatures ranging from 30°C to 80°C and are summarized in Table 9.

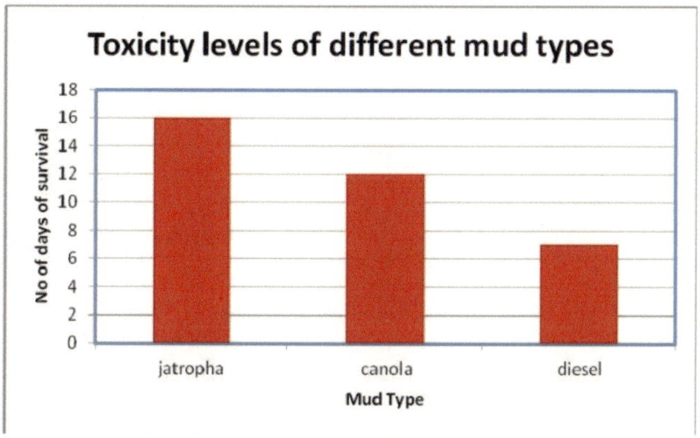

Figure 17. Toxicity of different mud types

Temperature	Diesel	Jatropha	Canola
30°C	10	10	10
40°C	10.1	10.05	10.05
50°C	10.17	10.1	10.05
60°C	10.2	10.15	10.1
70°C	10.2	10.15	10.15
80°C	10.25	10.2	10.17

Table 9. Density Changes in ppg at Varying Temperatures.

The mud samples were heated at constant pressure, and in an open system, hence the density increment.

At temperatures of 60°C and 70°C, the densities of Diesel and Jatropha OBM's were constant, while that happened with Canola OBM at a lower range of 40°C and 50°C. This is shown in Figure 18. This could be due to the differences in temperature and heat energy required to dissipate bonds, which vary with fluid properties (i.e the continuous phases).

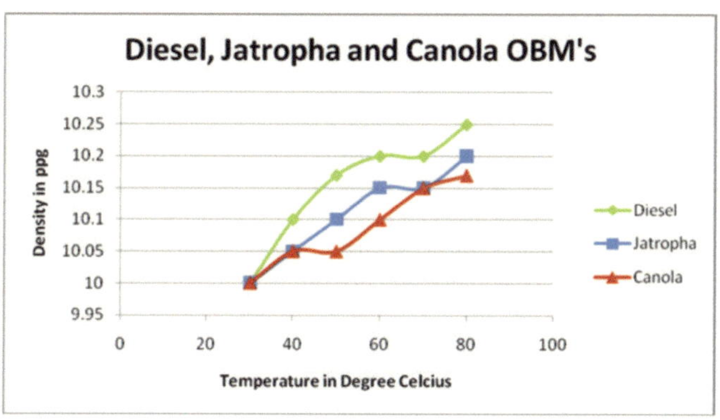

Figure 18. Density against Temperature (Diesel, Jatropha and Canola OBM's)

After the results were recorded, extrapolations were made and hypothetical values were derived for temperatures as high as 320°C, to enhance the prediction using Artificial Neural Network (ANN).

These values are summarized Tables 10 to 12

	Diesel	Jatropha	Canola
30°C	10	10	10
40°C	10.1	10.05	10.05
50°C	10.17	10.1	10.05
60°C	10.2	10.15	10.1
70°C	10.2	10.15	10.15
80°C	10.25	10.2	10.17
90°C	10.31133	10.24333	10.20667
100°C	10.35648	10.2819	10.24095
110°C	10.40162	10.32048	10.27524
120°C	10.44676	10.35905	10.30952
130°C	10.4919	10.39762	10.34381
140°C	10.53705	10.43619	10.3781
150°C	10.58219	10.47476	10.41238
160°C	10.62733	10.51333	10.44667
170°C	10.67248	10.5519	10.48095
180°C	10.71762	10.59048	10.51524
190°C	10.76276	10.62905	10.54952
200°C	10.8079	10.66762	10.58381
210°C	10.85305	10.70619	10.6181
220°C	10.89819	10.74476	10.65238
230°C	10.94333	10.78333	10.68667
240°C	10.98848	10.8219	10.72095
250°C	11.03362	10.86048	10.75524
260°C	11.07876	10.89905	10.78952
270°C	11.1239	10.93762	10.82381
280°C	11.16905	10.97619	10.8581
290°C	11.21419	11.01476	10.89238
300°C	11.25933	11.05333	10.92667
310°C	11.30448	11.0919	10.96095
320°C	11.34962	11.13048	10.99524

Table 10. Hypothetical Temperature-Density Values (extrapolated from regression analysis).

4.9. Results of neural networking

From the Artificial Neural Network Toolbox in the MATLAB 2008a, the following results were obtained:

60% of the data were used for training the network, 20% for testing, and another 20% for validation.

On training the regression values, returned values are summarized in Table 11

	Diesel	Jatropha	Canola
Training	0.99999	0.99999	0.99995
Testing	0.99725	0.99056	0.99898
Validation	0.99706	0.98201	0.99328
All	0.99852	0.99414	0.99675

Table 11. Regression Values.

Since all regression values are close to unity, this means that the network prediction is a successful one.

The graphs of training, testing and validation are presented below:

The values were returned after performing five iterations for each network. This also goes to say that the Artificial Neural Network, after being trained and simulated, is a viable and feasible instrument for prediction.

Figures 19 to 31 present the plots of Experimental data against Estimated (predicted) data for training, testing and validation processes from MATLAB 2008.

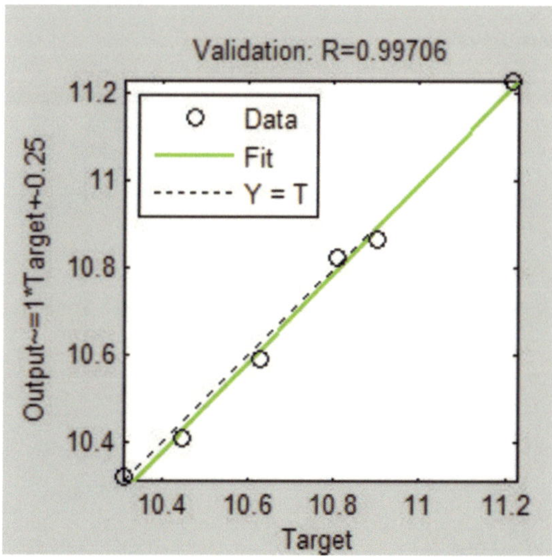

Figure 19. Diesel OBM Validation values

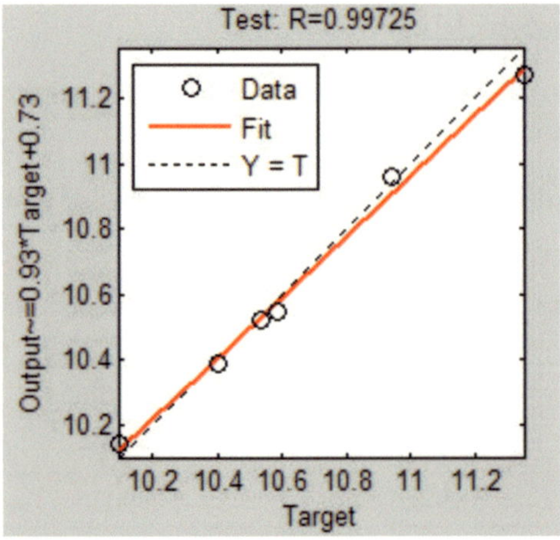

Figure 20. Diesel OBM Test values

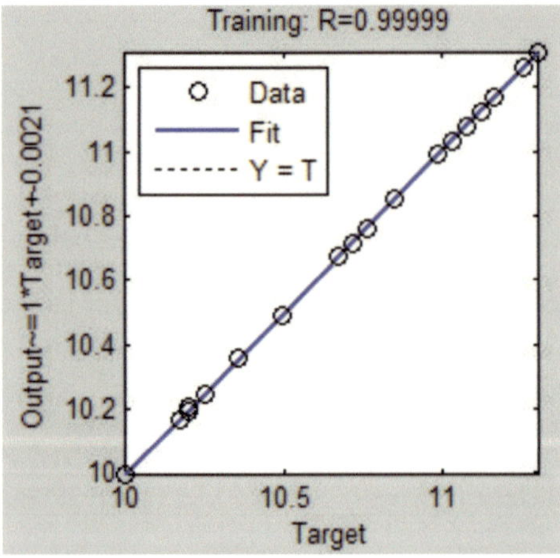

Figure 21. Diesel OBM Training values

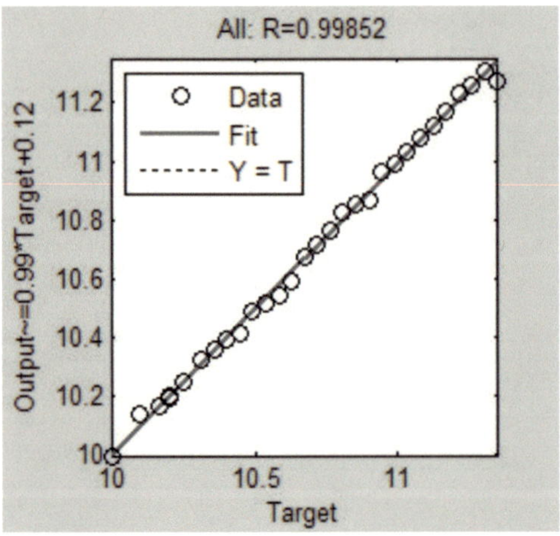

Figure 22. Diesel OBM Overall values

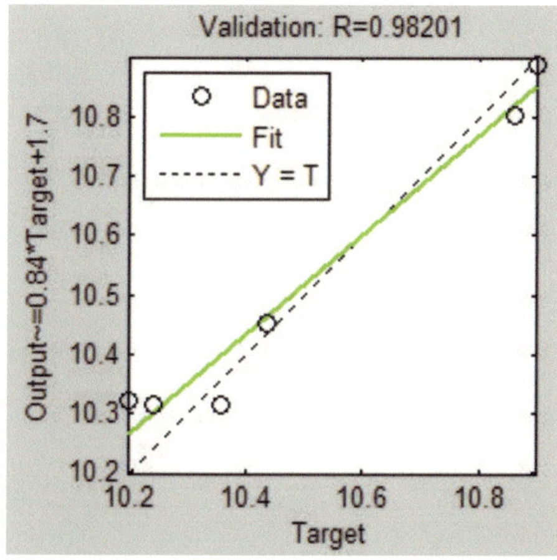

Figure 23. Diesel OBM Overall values

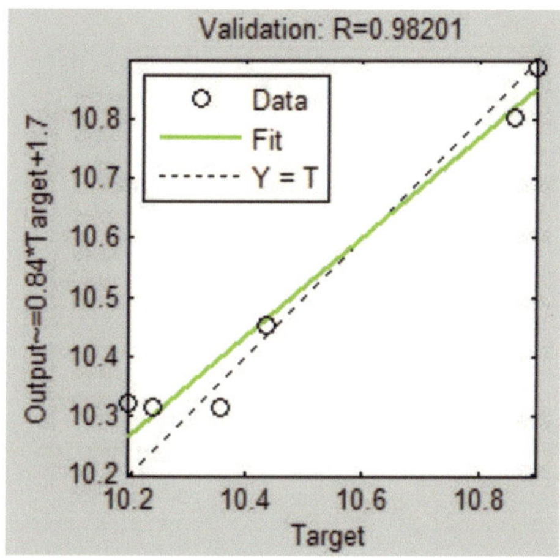

Figure 24. Jatropha OBM Validation values

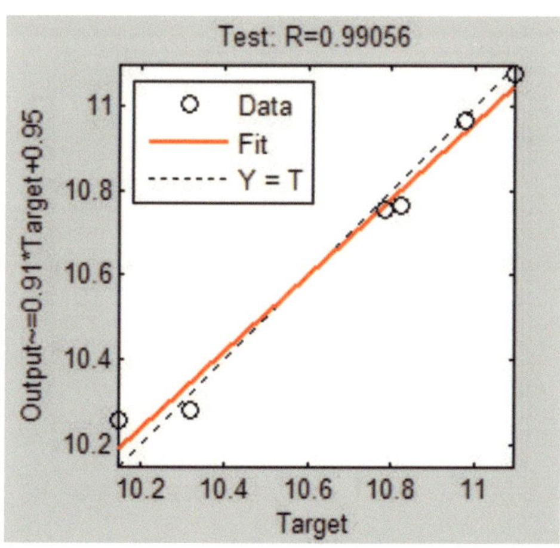

Figure 25. Jatropha OBM Test values

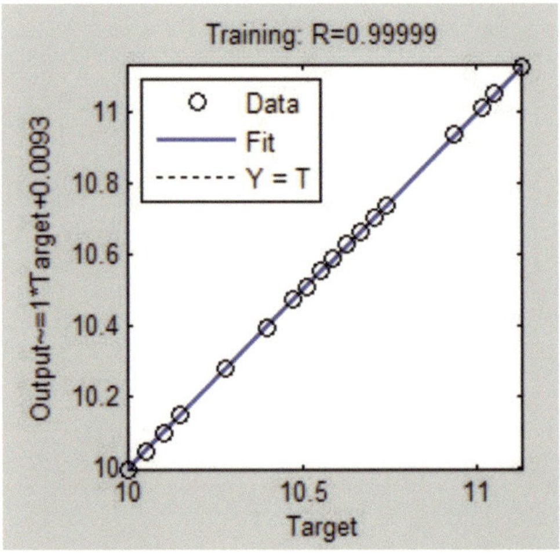

Figure 26. Jatropha OBM Training values

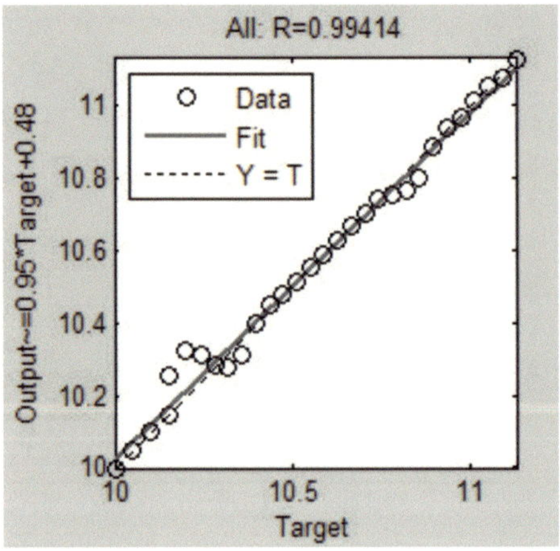

Figure 27. Jatropha OBM Overall values

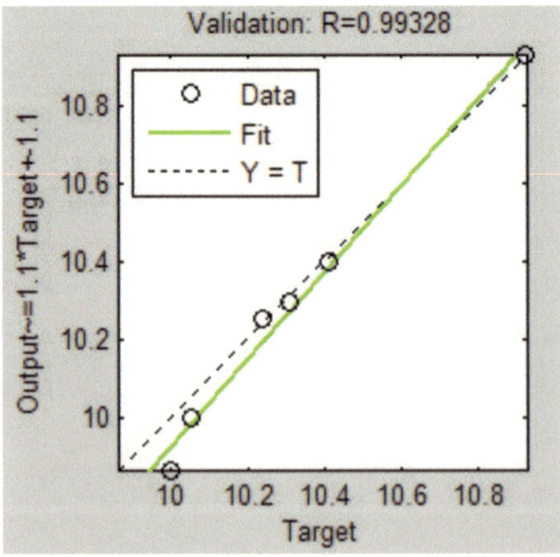

Figure 28. Canola OBM Validation values

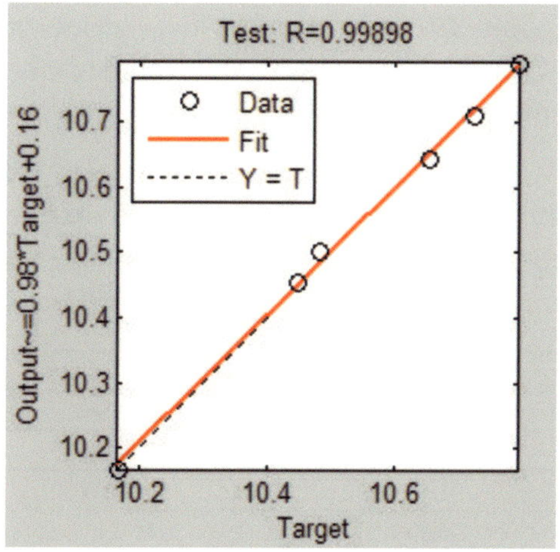

Figure 29. Canola OBM Test values

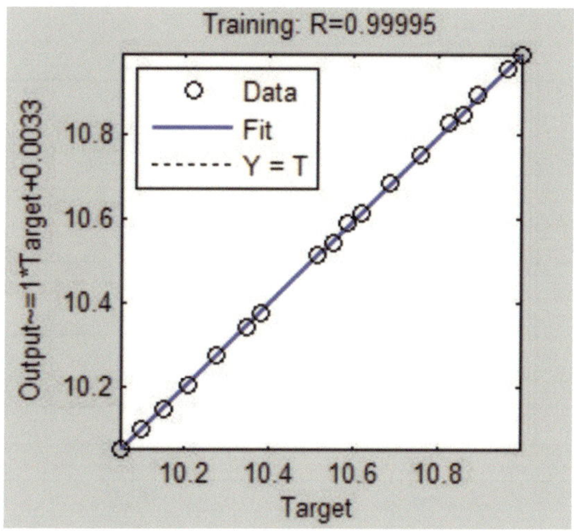

Figure 30. Canola OBM Training values

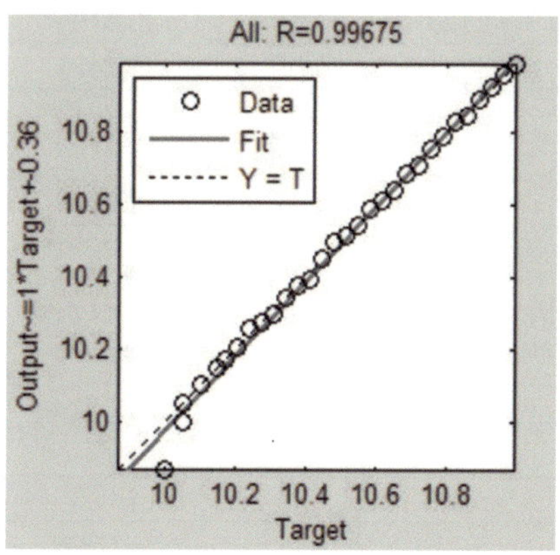

Figure 31. Canola OBM Overall values

We can see from the Figures 19 to 31 that the data points all align closely with the imaginary/arbitrary straight line drawn across. This validates the accuracy of the network predictions and this also gives rise to the high regression values (tending towards unity) presented in Table 11

Errors, estimated values and experimental values are summarized in Tables 12 to 14

Temperature °C	Exp Values	Est Values	Errors
30	10	10.049	0.049
40	10.1	10.1407	0.0407
50	10.17	10.1794	0.0094
60	10.2	10.2022	0.0022
70	10.2	10.2236	0.0236
80	10.25	10.24	-0.01
90	10.31133	10.287	-0.02433
100	10.35648	10.3579	0.001424
110	10.40162	10.3904	-0.01122
120	10.44676	10.4222	-0.02456
130	10.4919	10.4835	-0.0084
140	10.53705	10.5204	-0.01665
150	10.58219	10.5455	-0.03669
160	10.62733	10.6133	-0.01403
170	10.67248	10.687	0.014524
180	10.71762	10.7202	0.002581
190	10.76276	10.7714	0.008638
200	10.8079	10.8335	0.025595
210	10.85305	10.8611	0.008052
220	10.89819	10.8991	0.00091
230	10.94333	10.9623	0.018967
240	10.98848	10.9955	0.007024
250	11.03362	11.0273	-0.00632
260	11.07876	11.085	0.006238
270	11.1239	11.1195	-0.0044
280	11.16905	11.1474	-0.02165
290	11.21419	11.2049	-0.00929
300	11.25933	11.2432	-0.01613
310	11.30448	11.2545	-0.04998
320	11.34962	11.2674	-0.08222

Table 12. Errors, Experimental Values, and Estimated Values for Diesel OBM

Temperature °C	Exp Values	Est Values	Errors
30	10	10	0
40	10.05	10.05	0
50	10.1	10.0998	-0.0002
60	10.15	10.1485	-0.0015

Temperature °C	Exp Values	Est Values	Errors
70	10.15	10.2556	0.1056
80	10.2	10.3232	0.1232
90	10.24333	10.3143	0.070967
100	10.2819	10.2851	0.003195
110	10.32048	10.281	-0.03948
120	10.35905	10.3147	-0.04435
130	10.39762	10.3985	0.000881
140	10.43619	10.4526	0.01641
150	10.47476	10.4769	0.002138
160	10.51333	10.5126	-0.00073
170	10.5519	10.5544	0.002495
180	10.59048	10.5884	-0.00208
190	10.62905	10.63	0.000952
200	10.66762	10.6665	-0.00112
210	10.70619	10.7025	-0.00369
220	10.74476	10.741	-0.00376
230	10.78333	10.7559	-0.02743
240	10.8219	10.7655	-0.0564
250	10.86048	10.803	-0.05748
260	10.89905	10.8872	-0.01185
270	10.93762	10.9375	-0.00012
280	10.97619	10.9644	-0.01179
290	11.01476	11.0148	3.81E-05
300	11.05333	11.0533	-3.3E-05
310	11.0919	11.0747	-0.0172
320	11.13048	11.1305	2.38E-05

Table 13. Errors, Experimental Values, and Estimated Values for Jatropha OBM

Temperature °C	Exp Values	Est Values	Errors
30	10	9.8841	-0.1159
40	10.05	10.0044	-0.0456
50	10.05	10.048	-0.002
60	10.1	10.0925	-0.0075
70	10.15	10.1449	-0.0051
80	10.17	10.1681	-0.0019
90	10.20667	10.1987	-0.00797
100	10.24095	10.2489	0.007948
110	10.27524	10.2745	-0.00074
120	10.30952	10.2972	-0.01232
130	10.34381	10.3445	0.00069
140	10.3781	10.377	-0.0011
150	10.41238	10.4003	-0.01208
160	10.44667	10.4539	0.007233

Temperature °C	Exp Values	Est Values	Errors
170	10.48095	10.4994	0.018448
180	10.51524	10.519	0.003762
190	10.54952	10.5537	0.004176
200	10.58381	10.5952	0.01139
210	10.6181	10.6145	-0.0036
220	10.65238	10.6444	-0.00798
230	10.68667	10.6888	0.002133
240	10.72095	10.7105	-0.01045
250	10.75524	10.7365	-0.01874
260	10.78952	10.7895	-2.4E-05
270	10.82381	10.8224	-0.00141
280	10.8581	10.8465	-0.0116
290	10.89238	10.8971	0.004719
300	10.92667	10.9337	0.007033
310	10.96095	10.945	-0.01595
320	10.99524	10.9562	-0.03904

Table 14. Errors, Experimental Values, and Estimated Values for Canola OBM

The minute errors encountered in the predictions further justify the claim that the ANN is a trust worthy prediction tool.

The Experimental outputs were then plotted against their corresponding temperature values, and also fitted into the polynomial trend line of order 2.

The Equations derived are[7]:

Diesel OBM:

$$\rho = -4 \times 10^{-7} T^2 + 0.004T + 9.915 \quad (1)$$

Jatropha OBM:

$$\rho = 7 \times 10^{-7} T^2 + 0.003T + 9.994 \quad (2)$$

Canola OBM:

$$\rho = -2 \times 10^{-6} T^2 + 0.004T + 9.827 \quad (3)$$

Also by comparing the networks created with that of Osman and Aggour[12] (2003), we can see that this work is technically viable in predicting mud densities at varying temperatures as the network developed in the course of this project showed regression values close to those proposed by Osman and Aggour[12].

Errors, percentage errors and average errors as compared with Osman and Aggour[12] are relatively lower, thus guaranteeing the accuracy of the newly modeled network.

Table 15 shows the regression values of Osman and Aggour for oil based mud density variations with temperature and pressure[12].

Training	Testing	Validation	All
0.99978	0.99962	0.99979	0.9998

Table 15. Table Showing the Regression Values from Osman and Aggour[12]

Temperature	Diesel	Jatropha	Canola
30	0.49	0	1.159
40	0.40297	0	0.453731
50	0.092429	0.00198	0.0199
60	0.021569	0.014778	0.074257
70	0.231373	1.040394	0.050246
80	0.097561	1.207843	0.018682
90	0.235986	0.692808	0.078054
100	0.013748	0.031076	0.077606
110	0.107859	0.382504	0.007183
120	0.235115	0.428105	0.119538
130	0.080107	0.008473	0.006675
140	0.157991	0.157237	0.010553
150	0.346719	0.020412	0.116025
160	0.132049	0.006975	0.069241
170	0.136087	0.023647	0.176011
180	0.024081	0.019604	0.035776
190	0.080259	0.00896	0.039587
200	0.23682	0.01049	0.107622
210	0.074195	0.03447	0.03386
220	0.008346	0.035012	0.074922
230	0.173317	0.254405	0.019963
240	0.06392	0.521209	0.097495
250	0.057271	0.529223	0.174223
260	0.056307	0.108703	0.000221
270	0.039597	0.001088	0.013022
280	0.193818	0.107419	0.106789
290	0.082846	0.000346	0.043324
300	0.143289	0.000302	0.064369
310	0.442092	0.155111	0.145538
320	0.724421	0.000214	0.355045

Table 16. Table of the Relative Deviations

Table 17 compares the Average Absolute Percent Error abbreviation (AAPE), Maximum Average relative deviation (Ei) and Minimum E_i for Diesel, Jatropha and Canola OBM's as well as the values from Osman and Aggour.

	Diesel	Jatropha	Canola	Osman et al
Minimum E_i	0.008346	0.000214	0.000221	0.102269
Maximum E_i	0.724421	1.207834	1.159	1.221067
AAPE	0.172738	0.193426	0.124949	0.36037

Table 17. Table Comparing Maximum Ei, Minimum Ei, and AAPE

5. Conclusion

The lower viscosities of jatropha, moringa and canola oil based mud (OBM's) make them very attractive prospects in drilling activities.

The results of the tests carried out indicate that jatropha, moringa and canola OBM's have great chances of being among the technically viable replacements of diesel OBM's. The results also show that additive chemistry must be employed in the mud formulation, to make them more technically feasible. In addition, the following conclusions were drawn:

1. From the viscosity test results, it can be inferred that the plastic viscosity of jatropha OBM can be further stepped down by adding an adequate concentration of thinner. This method can also be used to reduce the gel strengths of jatropha, moringa and canola OBM's.
2. The formulated drilling fluids exhibited Bingham plastic behavior, and from the pressure loss modeling, canola OBM gave the best results, and next was jatropha OBM.
3. The tests of temperature effects on density: The densities increased and became constant at some point, and began increasing again (these temperature points of constant density varied for the different samples). The diesel OBM showed the highest variation range, while the canola OBM showed the lowest.
4. Artificial Neural Network works well for prediction of scientific parameters, due to minimized errors returned.

6. Limitations

1. The temperature-density tests were carried out at surface conditions under an open system and at a constant pressure due to the absence of a pressure unit thus, the equations developed are not guaranteed for down-hole circulating conditions.
2. During the temperature-density tests, it was observed that some of the mud particles settled at the base of the containing vessel, and this reduced the accuracy of the readings.
3. The accuracy of the temperature-density readings is also reduced because of the use of an analogue mud balance (calibrated to the nearest 0.1 ppg).
4. The mud samples were aged for only 24 hours, hence the feasibility of older muds may not be guaranteed.

7. Recommendations

1. This work should further be tested and investigated for the effect of temperature on other properties of the formulated drilling fluids.

2. The temperature-density tests should also be carried out at varying pressures, to simulate downhole conditions.

Author details

Adesina Fadairo, Churchill Ako, Abiodun Adeyemi and Anthony Ameloko
Department of Petroleum Engineering, Covenant University, Ota, Nigeria

Olugbenga Falode
Department of Petroleum Engineering, University of Ibadan, Nigeria

Acknowledgement

We wish to thank all members of staff Department of Petroleum Engineering Covenant University, Nigeria for their technical support in carrying out this research work especially Mr Daramola. We also acknowledge the support of Environmental Research Group, Father-Heroes Forte Technology Nigeria for their commitment.

8. References

[1] Yassin .A., Kamis .A., Mohamad .O.A., (1991) *"Formulation of an Environmentally Safe Oil Based Drilling Fluid"* SPE 23001, Paper presented at the SPE Asia Pacific Conference held in Perth, Western Australia, 4-7 November 1991.

[2] Terry Hemphil, (1996)*"Prediction of Rheological Behavior of Ester-Based Drilling Fluids Under Downhole Conditions"* SPE 35330. Paper presented at .s1 Ihe 1996 SPE International Petroleum Conference and Exhibition of Mexico held in Villahermosa, Tabasco 5-7 March 1996.

[3] A.M. Ezzat and K.A. Al-Buraik, (1997) *"Environmentally Acceptable Drilling Fluids for Offshore Saudi Arabia"* SPE 37718. Paper presented at the SPE Middle East Oil Show and Conference held in Bahrain, 15-18 March, 1997.

[4] G. Sáchez; N. León, M. Esclapés; I. Galindo; A. Martínez; J. Bruzual; I. Siegert, (1999) *"Environmentally Safe Oil-Based Fluids for Drilling Activities"* SPE 52739, Paper presented at SPE/EPA Exploration and Production Environmental Conference held in Austin, Texas, 28 February–3 March 1999.

[5] Xiaoqing .H. and Lihui .Z., (2009) *"Research on the Application of Environment Acceptable Natural Macromolecule Based Drilling Fluids"* SPE 123232, Paper presented at the SPE Asia Pacific Health, Safety, Security and Environment Conference and Exhibition held in Jarkata, Indonesia, 4-6 August 2009.

[6] Dosunmu .A. and Ogunrinde .J. (2010) *"Development of Environmentally Friendly Oil Based Mud using Palm Oil and Groundnut Oil"*. SPE 140720. Paper presented at the 34th Annual International Conference and Exhibition in Tinapa-Calabar, Nigeria, July 31st-August 7th, 2010.

[7] *Fadairo Adesina, Adeyemi Abiodun, Ameloko Anthony, Falode Olugbenga, (SLO 2012) "Modelling the Effect of Temperature on Environmentally Safe Oil Based Drilling

Mud Using Artificial Neural Network Algorithm" *Journal of Petroleum and Coal 2012, Volume 54, Issue 1.*

[8] *Fadairo Adesina, Ameloko Anthony, Adeyemi Gbadegesin, Ogidigbo Esseoghene, Airende Oyakhire (2012) "Environmental Impact Evaluation of a Safe Drilling Mud" *SPE Middle East Health, Safety, Security, and Environment Conference and Exhibition held in Abu Dhabi, UAE, 2–4 April 2012,* SPE-152865-PP

[9] Bourgoyne .T.A., Millheim .K.K, Chenevert M.E., *"Applied Drilling Engineering"*. SPE Textbook series, Vol 2, 1991.

[10] Mitchell B., *Advanced Oil Well Drilling Engineering Hand Book,* Mitchell Engineering, 10th edition, 1995.

[11] Baker Hughes Mud Engineering Hand Book

[12] Osman, E.A. and Aggour, M.A.: "Determination of Drilling Mud Density Change with Pressure and Temperature Made Simple and Accurate by ANN," paper SPE 81422 presented at the 2003 SPE Middle East Oil Show and Conference, Bahrain, 5-8 April.

Section 2

IT and Modeling

Chapter 4

Alternative Computing Platforms to Implement the Seismic Migration

Sergio Abreo, Carlos Fajardo, William Salamanca and Ana Ramirez

Additional information is available at the end of the chapter

http://dx.doi.org/10.5772/51234

1. Introduction

High performance computing (HPC) fails to satisfy the computational requirements of Seismic Migration (SM) algorithms, due to problems related to computing, the data transference and management ([4]). Therefore, the execution time of these algorithms in state-of-the-art clusters still remain in the order of months [1]. Since SM is one of the standard data processing techniques used in seismic exploration, because it generates an accurate image of the subsurface, oil companies are particularly interested in reducing execution times of SM algorithms.

Computational migration needed for large datasets acquired today is extremely demanding, and therefore the computation requires CPU clusters. The performance of CPU clusters had been duplicating each 24 months until 2000 (satisfying the SM demands), but since 2001 this technology stop accelerating due to three technological limitations known as Power Wall, Memory Wall and IPL Wall [10, 23]. This encouraged experts all over the world to find new computing alternatives.

The devices that have been highlighted as a base for the alternative computing platforms are FPGAs and GPGPUs. These technologies are subject of major research in HPC since they have a better perspective of computing [10, 14]. Different implementations of SM algorithms have been developed using those alternatives platforms [4, 12, 17, 21, 22, 29]. Results show a reduction in the running times of SM algorithms, leading to combine these alternative platforms with traditional CPU clusters in order to get a promising future in HPC for seismic exploration.

This chapter gives an overview of the HPC with an historical perspective, emphasizing in the technological aspects related to the SM. In the section two we will show the seismic migration evolution together with the technology, section three summarizes the most important aspects of the CPU operation in order to understand the three technological walls, section four presents the use of GPUs as a new HPC platform to implement the SM, section five shows the use of FPGAs as another HPC platform, section six discusses the advantages and disadvantages of both technologies and, finally, the chapter is closed with the acknowledgments.

At the end of this chapter,it is expected that the reader have enough background to compare platform specifications and be able to choose the most suitable platform for the implementation of the SM.

2. The technology behind the Seismic Migration[1]

One of the first works related with the beginnings of the Seismic Migration (SM) is from 1914 and it was related with the construction of a sonic reflection equipment to locate icebergs by the Canadian inventor Reginald Fessenden. The importance of this work was that together with the patent named "Methods and Apparatus for Locating Ore Bodies" presented by himself three years later (1917), he gave us the first guidelines about the use of reflection and refraction methods to locate geological formations.

Later, in 1919, McCollum and J.G. Karchner made other important advance in this field because they *received a patent for determining the contour of subsurface strata that was inspired by their work on detection of the blast waves of artillery during World War I*,[6]. But it was only in 1924 when a group led by the German scientist Ludger Mintrop, could locate the first Orchard salt dome in Fort Bend County, Texas [15].

In the next year (1925), as a consequence of the Reginald's patent in 1917, Geophysical research corporation (GRC) created a department dedicated to the design and construction of new and improve seismographic instrumentation tools.

2.1. Acceptance period

In the next three years (1926 - 1928) GRC was testing and adjusting his new tools, but at that time there was an air of skepticism due to the low reliability of the instruments. It can be said that the method was tested and many experiments were performed, but it was only between 1929 and 1932 when the method was finally accepted by the oil companies.

Subsequently, the oil companies began to invest large sums of money to improve it and have a technological advantage over their competitors. *As the validity of reflection seismics became more and more acceptable and its usage increased, doubts seemed to reenter the picture. Even though this newfangled method appeared to work, in many places it produced extremely poor seismic records. These were the so-called no-record areas* [6].

Only until 1936 when Frank Reiber could recognize that the cause of this behavior was the steep folding, [38], faulting or synclinale and anticlinal responses and he built an analog device to model the waves of different geological strata. This discovery was really important for the method, because it could give it the solidity that the method was needing at that moment; but finally the bad news arrived with the beginning of the world war II, because it stopped all the inertia of this process.

[1] The main ideas of this section has been taken from the work of J. Bee Bednar in his paper A brief history of seismic migration

2.2. World War II (1939-1945)

With the World War II all efforts were focused on building war machines. During this period, important developments were achieved, which would be very useful in the future of the SM. Some of these developments were done by Herman Wold, Norbert Weiner, Claude Shannon and Norman Levinson from MIT, and established the fundamentals of the numerical analysis and finite differences, areas that would be very important in future of seismic imaging. For example, Shannon proposed a theorem to sample analog signals and convert them into discrete signals and then developed all the mathematical processing of discrete signals starting the digital revolution.

On the other hand, between 1940 and 1956, appeared the first generation of computers which used Vacuum Tubes (VT). The VT is an electronic device that can be used as an electrical switch to implement logical and arithmetic operations. As the VT were large and consumed too much energy, the first computers were huge and had high maintenance and operation costs. They used magnetic drums for memory, their language was machine language (the lowest level programming language understood by computers) and the information was introduced into them through punched cards[2] [36].

One of the first computers was developed in Pennsylvania University, in 1941 by John Mauchly and J. Prester and it was called **ENIAC** [36]. This computer had the capacity of performing 5000 additions and 300 multiplications per second (Nowadays, a PC like Intel core i7 980 XE can perform 109 billions of Floating Point Operations [27]). In the next years were developed the **EDVAC** (1949), the first commercial computer **UNIVAC** (1951) [36].

One final important event on this period, was the general recognition that the subsurface geological layers weren't completely flat (based on Rieber work previously mentioned), leading the oil companies to develop the necessary mathematical algorithms to locate correctly the position of the reflections and in this way strengthen the technique.

2.3. Second generation (1956-1963): Transistors

In this generation the VT were replaced by transistors (invented in 1947). The transistor also works as an electric switch but it's smaller and consumes less energy than the VT (see figure 1). This brought a great advantage for the computers because made them smaller, faster, cheaper and more efficient in energy consumption.

Additionally this generation of computers started to use a symbolic language called assembler (see figure 2) and it was developed the first version of high level languages like COBOL and FORTRAN. This generation also kept using the same input method (punched cards) but changed the storage technology from magnetic drum to magnetic core.

On the other hand, the SM in this period received a great contribution with the J.G. Hagedoorn work called "A process of seismic reflection interpretation" [18]. In this work Hagedoorn introduced a "ruler-and-compass method for finding reflections as an envelope of equal traveltime curves" based on Christiaan Huygens principle.

Other important aspect in this period was that the computational technology began to be used in seismic data processing, like the implementation made by Texas Instrument Company in

[2] A punched card, is an input method to introduce information, through the hole identification

Figure 1. In order from left to right: three vacuum tubes, one transistor, two integrated circuits and one integrated microprocessor.

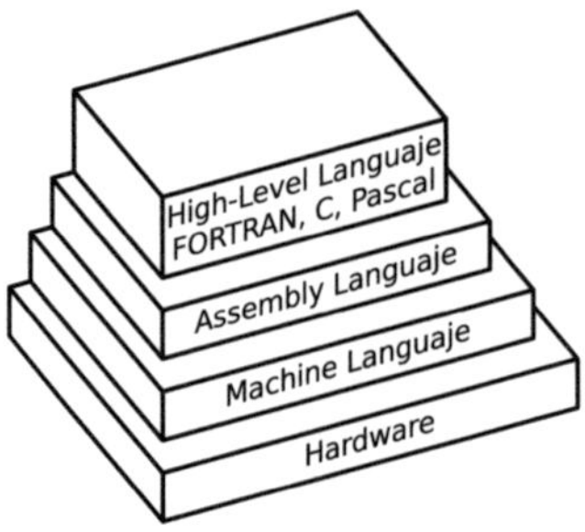

Figure 2. Assembly language.

1959 on the computer TIAC. In the next year (1960) Alan Trorey from Chevron Company developed one of the first computerized methods based on Kirchhoff and in 1962 Harry Mayne obtained a patent on the CMP stack [30]; this two major contributions would facilitate later the full computational implementation of the SM.

2.4. Third generation (1964-1971): Integrated circuits

In this generation the transistors were miniaturized and packaged in integrated circuits (IC) (see figure 1). Initially the IC could contain less than 100 transistors but nowadays they can contain billions of transistors. With this change of technology, the new computers could be faster, smaller, cheaper and more efficient in the consumption of energy than the second generation [36].

Additionally, the way in which the user could introduce the information to the computers also changed, because the punched cars were replaced by the monitor, keyboard and interfaces based on operating systems (OS). The OS concept appeared for first time, allowing to this new generation of computers execute more than one task at the same time.

On the other hand, the SM also had a significant progress in this period. *In 1967, Sherwood completed the development of Continuous Automatic Migration (CAM) on an IBM accounting machine in San Francisco. The digital age may have been in its infancy, but there was no question that it was now running full blast,* [6] and, in 1970 and 1971, Jon Claerbout published two papers focused on the use of second order, hyperbolic, partial-differential equations to perform the imaging. Largely, the Claerbout work was centered in the use of finite differences taking advantage of the numeric analysis created during the World War II. The differences finite work, allowed to apply all these developments over the computers of that time.

2.5. Fourth generation (1971- Present): Microprocessors

The microprocessor is an IC that works as a data processing unit, providing the control of the calculations. It could be seen as the computer brain, [36].

This generation was highlighted because each 24 months the number of transistors inside an IC was doubled according with the Moore Law, who predicted this growing rate in 1975, [32]. This allowed that the development of computers with high processing capacity were growing very fast (see table 1). In 1981 IBM produced the first computer for home, in 1984 Apple introduced the Macintosh and in 1985 was created the first domain on Internet (symbolics.com). During the next years the computers were reducing their size and cost; and taking advantage of Internet, they raided in many areas.

Microprocessors	Year.	Number of transistors	Frequency	bits
Intel 4004	1971	2,300	700 KHz	4
Intel 8008	1972	3,300	800 KHz	8
Intel 8080	1974	4,500	2MHz	8
Intel 8086	1978	29,000	4MHz	16
Intel 80386	1980	275,000	40MHz	32
Intel Pentium 4	2000	42,000,000	3.8GHz	32
Intel Core i7	2008	731,000,000	3GHz	64

Table 1. Summary of the microprocessors evolution.

Moore law was in force approximately until 2000, because of power dissipation problems produced by the amount of transistors inside a single chip, this effect is known as the Power Wall (PW) and remains stagnating the processing capacity.

From that moment began to surface new ideas to avoid this problem. Ideas like fabricate processors with more than one processing data unit (see figure 3); these units are known as cores, seemed to be a solution but new problems arose again. Problems like the increase of cost, power consumption and design complexity of the new processors, didn't reveal to be a good solution. This is the reason that even today we can see that the processors fabricated with this idea only have got 2, 4, 6, 8, or in the best way 16 cores, [5].

In figure 4 we can see the Moore Law tendency until 2000, and after this date we can see the emergence of multi-core processors.

2.6. HPC's birth

A new challenge to computing researchers was the implementation of computationally expensive algorithms. For this purpose was necessary to design new strategies to do an optimal implementation over the more powerful computers. That group of implementation strategies over supercomputers has been known as High Performance Computing (HPC).

The HPC was born in Control Data Corporation (CDC) in 1960 with Seymour Cray, who launched in 1964 the first supercomputer called CDC 6600 [24]. Later in 1972 Cray left CDC

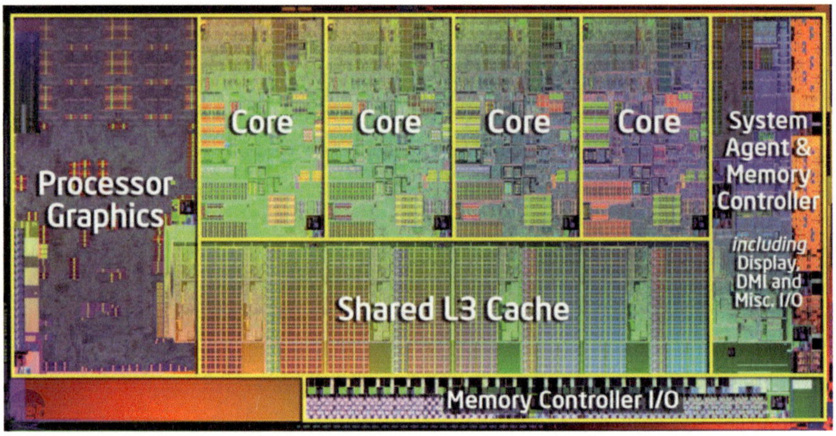

Figure 3. Intel multi-core processor die. Taken from [8]

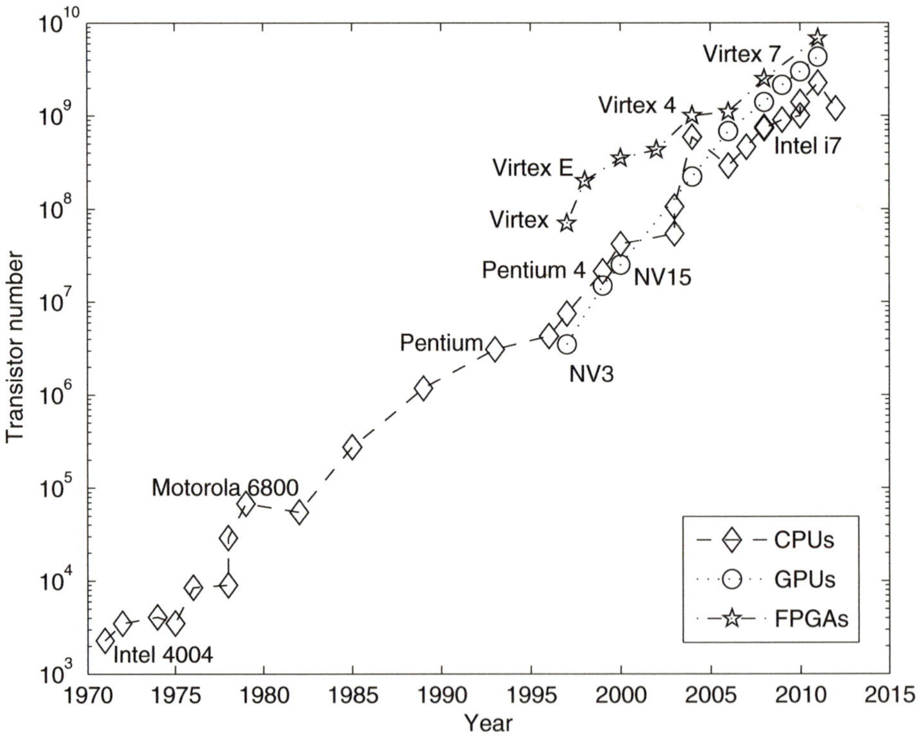

Figure 4. CPU, GPU and FPGA Transistor Counts

to create his own company and four years later in 1976, Cray developed Cray1 which worked at 80MHz and it became in the most famous supercomputer of the history.

After that, the HPC kept working with clusters of computers using the best cores (processors) of each year. Each cluster was composed by a front-end station and nodes interconnected through the Ethernet port (see figure 5). So, in that way the HPC evolution was focused on increase the quantity of nodes per cluster, improve the interconnection network and the development of new software tools that allow to take advantage of the cluster. This strategy began to be used by other companies in the world like Fujitsu's Numerical Wind Tunnel, Hitachi SR2201, Intel Paragon among others, giving very good results.

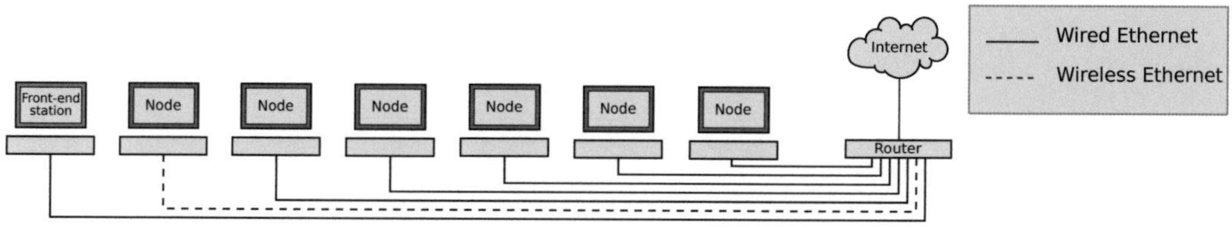

Figure 5. General diagram of a cluster

2.7. SM consolidation

On the other hand, the SM also received a great contribution by Jon Claerbout, because in 1973 he formed The Stanford Exploration Project (SEP). The SEP together with the group GAC from MIT built the bases for many consortia that would be formed in the future like Center For Wave Phenomenon at Colorado School of Mines. In 1978 R. H. Stolt presented a different approach to do the SM, which was called "Migration by Fourier Transform", [41]. Compared with the approach of Claerbout, the Stolt migration is faster, can handle up to 90 degrees formation dips but it's limited to constant velocities; while as Claerbout migration is insensitive to velocity variations and can handle formation dips up to 15 degrees.

In 1983 three papers were published at the same time about a new two-way solution to migration; these works were done by Whitmore, McMechan, and Baysal, Sherwood and Kosloff. Additionally, in 1987 Bleistein published a great paper about Kirchhoff migration, using a vastly improved approach to seismic amplitudes and phases. From that moment, the SM could be done in space time (x,t), frequency space (f,x), wavenumber time (k,t), or almost any combination of these domains.

Once the mathematical principles of the SM were developed, the following efforts focused on propose efficient algorithms. The table 2 summarizes some of these implementations, [6].

But perhaps one of the most important software developments between 1989 and 2004 for seismic data processing was performed by Jack Cohen and John Stockwell at the Center for Wave Phenomena (CWP) at the Colorado School of Mines. This tool was called Seismic Unix (SU) and has been one of the most used and downloaded worldwide.

Moreover, the combination of cluster of computers with efficient algorithms began to bear fruit quickly, allowing pre-stack migrations in depth at reasonable times. An important aspect to note is that almost all algorithms that we use today were developed during that time and have been implemented as the technology has been allowing it.

Year	Author	Contribution
1971	W. A. Schneider's	tied diffraction stacking and Kirchhoff migration together.
1974 and 1975	Gardner et al	Clarified this even more.
1978	Schneider	Integral formulation of migration.
1978	Gazdag	Phase shift method.
1984	Gazdag and Sguazzero	Phase shift plus interpolation.
1988	Forel and Gardner	A completely velocity-independent migration technique.
1990	Stoffa	The split step method.

Table 2. Development of efficient algorithms.

Over the years the seismic imaging began to be more demanding with the technology, because every time it would need a more accurate subsurface image. This led to the use of more complex mathematical attributes. Additionally it became necessary to process a large volume of data resulting from the 3D and 4D seismic and the use of multicomponent geophones. All this combined with the technological limitations of computers since 2000, made HPC developers began to seek new technological alternatives.

These alternatives should provide a solution to the three barriers currently faced by computers (Power wall, Memory wall and IPL wall; these three barriers will be explained in detail below). For that reason the attention was focused on two emerging technologies that promised to address the three barriers in a better way than computers, [14]. These technologies were GPUs and FPGAs.

In the following sections it's explained how these two technologies have been facing the three barriers, in order to map out the future of the HPC in the area of seismic data processing.

3. Understanding the CPUs

In this section initially we will present to the reader an idea about the internal operation of a general purpose processing unit (CPU), assuming that the reader doesn't have any previous knowledge. Subsequently, we will analyze the contribution made by the alternative computing platforms in the high performance computing.

Let's imagine an automobile assembly line completely automated like the shown in the figure 6, which produce type A automobiles. The parts of the cars can be found in the warehouse and are taken into the line, passing through the following stages: engine assembly, bodywork assembly, motor drop, painting and quality tests. Finally we get the vehicle and it is carried to the parking zone. In this way, in the seismic data processing, a computing system can access the data stored in memory, adapt it, calculate the velocity model, and finally migrate them to produce a subsurface image that is stored again in memory.

Now let's suppose that we want to modify the assembly line in order to make it produce type B automobiles. Therefore, it's necessary either to add some stages to the assembly line or to modify the existing modules configuration. In like manner operates a CPU, where it can be executed different kind of algorithms as well as the assembly line can now produce different

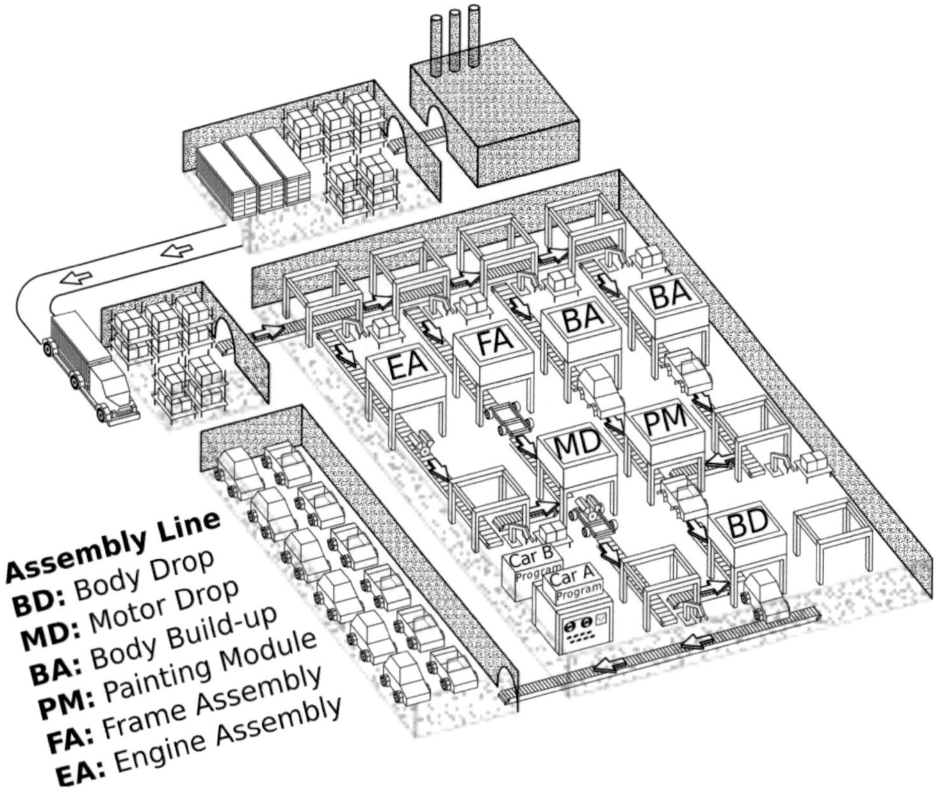

Figure 6. CPU analogy

kinds of automobiles. A CPU includes different kind of modules that allow it execute several instructions which can execute the desired algorithm.

As well as the CPU receive and deliver data, the assembly line receive parts and deliver vehicles. But to assemble a particular type of car, the line have to get prepared. In this way, the line can now select the right parts from the warehouse, transport it and handle it suitably. In a CPU, the program is the responsible to indicate how to get each data and how to manipulate it to obtain the final result.

A program is composed by elementary process called instructions. In the assembly line, an instruction matched simple actions like bring a part from the warehouse, adjust a screw, activate the painting machine, etc. In a CPU, an instruction executes a elementary calculation over the data such as fetch data from memory, carry out a sum, multiplication, etc.

3.1. CPU operation

In a CPU, the instructions are stored in the memory and are executed sequentially one by one as it's shown in the figure 6. To execute one instruction, it must fetch it from memory, decode (interpret) it, read the data required by the instruction, manipulate the data and finally return the results to the memory. This architecture is known as Von Neumann. Moderns CPUs have evolved from their original concept, but its operating principle is still the same.

Nowadays almost any electronic component that we use is based on CPUs like our PC, cell phones, videogames, electrical appliance, automobiles, digital clocks, etc.

As was described, the CPU is in charge to process the data stored in memory as the program indicated. In our assembly line we can identify several blocks that handle, transport and fit the vehicle parts. In a CPU, this set of parts is known as the datapath. The main functionality of the datapath is to temporally store, transform and route the data in a track from the entrance to the exit.

In the same way, in the assembly line we can find the main controller which is in charge to monitor all the process and to activate harmonically the units on each stage to execute the assembly process. All this, taking into account the requested assembly process. In a CPU we can find the control unit, in charge to address orderly the data through the functional units to execute each one of the instructions that the program indicates and thus perform the algorithm.

3.2. CPU performance

The performance of an assembly line, could be measured as the time required to produce certain amount of vehicles. In the same way, the CPU performance is measured as how long it takes to process some data. In first place, the performance of the assembly line could be limited by the speed of its machines as well as in the CPU the integrated circuit speed is proportional to its performance. The second aspect that affects the performance is how fast can be put the parts at the beginning of the assembly line. In the same way, the data transference rate between CPU and memory could limit its performance. The CPU performance is related with the execution time of an algorithm, for that reason we are going to analyze some aspects that have slowed down the CPU performance.

We will analyze the assembly line operation to make the best use of its machines, and increase its performance. We can observe that the units on each stage could simultaneously work over different vehicles, and it is not necessary to finish one vehicle to start the next one (see figure 6). In the same way, the CPU can process several data at the same time provided that it has been designed using the technique called pipeline, [19]. This technique segments the execution process and allows that the CPU handles several data in parallel. This is one of the digital techniques developed to improve the CPU speed although nowadays developments on this area are stuck. This phenomenon is one of the greatest performance improvement constraint and it's called the Instruction Level Parallelism (ILP) wall [23].

The ILP can be associated with a technique that tries to reorganize the assembly line machines, in order to improve the performance. Additionally, there have been other different ways to improve the performance, one of them is using new machines more efficient that can assemble the parts faster. Also these new machines could be smaller in size, occupying less space, allowing, therefore more machines in the same area, increasing the productivity. In the CPU context, it has been the improvement of the integrated circuit manufacturing which has allowed the deployment of smaller transistors. This supports the implementation of more dense systems (i.e. the greatest number of transistors per chip) and faster (i.e. that can perform more instructions per unit time).

The growth rate of the amount of transistors in integrated circuits has faithfully obeyed Moore's Law, it allows to have smaller transistors on a chip. As the transistors were becoming smaller, their capacitances were reducing allowing shorter charges and discharges times. All this allow higher working frequencies and there is a direct relationship between the working frequency of an integrated circuit and power dissipation (given by equation 1 taken from [9]).

$$P = C\rho f V_{dd}^2 \tag{1}$$

where ρ represents the transistor density, f is the working frequency of the chip, V_{dd} is the power supply voltage and C is the total capacitance. Nowadays the power dissipation has grown a lot and has become an unmanageable problem with the conventional cooling techniques. This limitation on the growth of the processors is known as the power wall and has removed one of the major growth forces of the CPUs.

3.3. Memory performance

Returning to our assembly line, we saw it with a lot of high speed machines, but now its performance could be limited by the second constraint of the system performance. The speed of the incoming parts to the assembly line could not be enough to keep all these machines busy. If we don't have all parts ready, we will be forced to stop until they become available again. In the same way, faster CPUs require more data ready to be processed. The limitation presented nowadays in the communication channels to supply the demand of data processing units is called memory wall.

This wall was initially treated by improving the access paths between the warehouse and the assembly line, which in a CPU, means improving the transmission lines between memory and CPU on the printed circuit board.

The second form, the memory wall was faced, is more complex, but takes into account all the process since the manufacturer deliver. Analyzing the features of the warehouse parts, we can find different kinds. The part manufacturers have large deposits with a lot of parts ready to be used, but use a part from these deposits is expensive in time. On the other hand, the warehouse next to the assembly line have stored a smaller number of pieces of different kinds, because we must remember that the assembly line can produce different types of cars. The advantage of this warehouse is that the parts reach faster the points of the production line where they are needed. Some deposits are larger but their transportation time to the line is slower, besides, the nearest deposit has a lower storage capacity. The line could improve its performance taking advantage of this deposit features.

Similarly occurs with the computing system memory. High capacity memory such as hard disks, are read by blocks, and access one single data requires to transport a large volume of information that is locally stored in RAM, which represents a nearest deposit. Then the single data can be read and taken to the CPU. The memory organization in a modern computing system is arranged hierarchically as the figure 6 shows. Even in this organization, the assembly line has provided small storage spaces next to the machines that require specific parts. In a CPU this is called CPU cache.

The memory hierarchy creates local copies of the data that will be handled in certain algorithm on fast memories to speed up its access. The challenge of such systems is always have available the required data in the fast memory, because otherwise, the CPU must stop while the data is accessed in the main memory.

Currently, the three computing barriers are present and they are the main cause of stagnation in which CPU based technology has fallen. Some solutions have been proposed based on alternative technologies, such as the graphics processing units (GPU) and Field Programmable Gate Array (FPGA), and have intended to mitigate these phenomena, [4, 22]. They seem to be a short and mid-term solutions respectively.

4. GPUs: A computing short term alternative

GPUs are the product of the gaming industry development that was born in the 50′ and started a continuous growing market on 80′. Video-games require to execute intensive computing algorithms to display the images, but unlike other applications such as seismic migration, the interaction with the user requires the execution in a very short time. Therefore, since the 90′ have been developed specialized processors for this purpose. These processors have been called GPU, and they have been widely commercialized in video-game consoles and PC video acceleration cards, [37].

GPUs are specific application processors that reach high performance on the task that they were designed for, in the same way that the assembly line would outperform itself if it were dedicated to assembly a little range of vehicle types. This is achieved because the unnecessary machines can be eliminated and the free area is optimized. Likewise improve the availability, storage and handling of the parts(See figure 7).

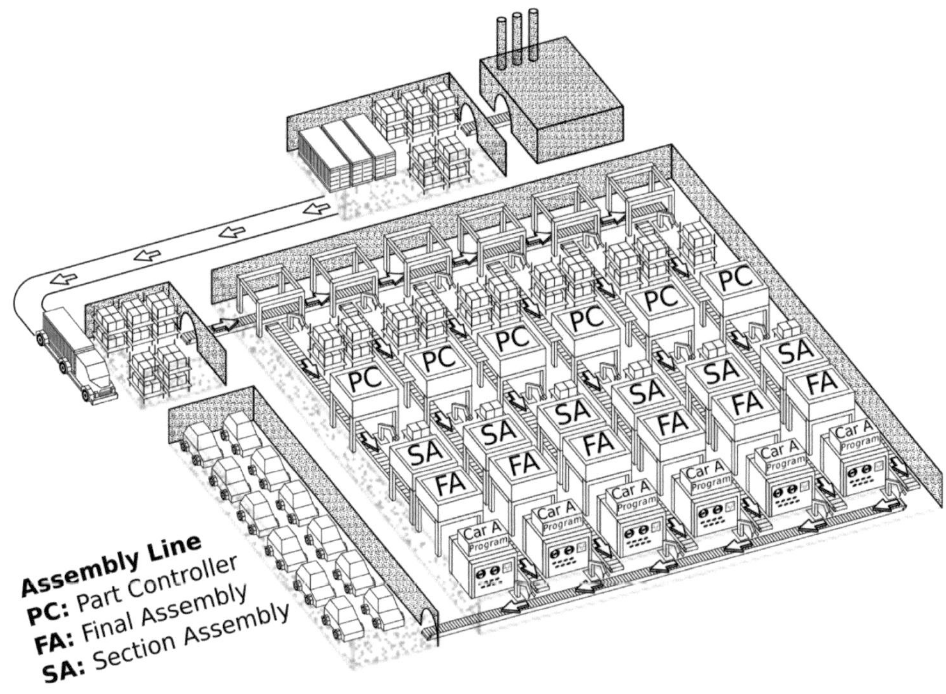

Figure 7. GPU analogy

The task of a GPU is an iterative process that generates an image pixel by pixel (point by point) from some data and input parameters. This allows to process in parallel each output pixel and therefore GPUs architecture is based on a set of processors that perform the same algorithm for different output pixels. It is very similar to the sum of the contributions made in Kirchhoff migration, so in that way this architecture is a good option to try to speed up a migration process.

The memory organization is another relevant feature of the GPU architecture. This allows all the processors to access common information to all pixels that are being calculated and in the same way each processor can access particular pixel information.

Like our assembly line, the GPU task can be segmented in several stages, elementary operations or specific graphics processing instructions. Initially GPUs could only be used in graphical applications and although their performance in these tasks was far superior than a CPU, its computing power could only be used to carry out this task. For this reason the GPU manufacturers made a first contribution in order to make them capable to execute any algorithm; they make their devices more flexible and have become General Purpose GPU (GPGPU).

This make possible to exploit the computing power of these devices in any algorithm, but it was not an easy task for programmers. Making an application required a deep understanding of the GPGPUs instructions and architecture, so programmers were not very captivated. Therefore, the second contribution that definitely facilitated the use of these devices was the development in 2006 of programming models that does not require advanced knowledge on GPUs such as Compute Unified Device Architecture (CUDA).

GPGPUs are definitely an evolved form of a CPU. Their memory management is highly efficient, which has reduced the memory wall; and because of the number of processing cores on a single chip, they have been forced to reduce the working speed and to operate at the limit of the power wall. Its growth rate seems to continue stable and it promises to be in a mid-term with the technology that drives the high performance computing. But these three barriers are still present and this technology soon will be faced them.

5. FPGAs: Designing custom processors

Another alternative to accelerate the SM process are the FPGAs (Field Programmable Gate Array), these devices are widely used in many problems of complex algorithms acceleration, for this reason, since a couple of years, some traditional manufacturers of high performance computing equipment began to include in their brochure, computer systems that include FPGAs.

To get an initial idea of this technology, let us imagine that the car assembly is going to be amended as follows:

- The assembly plant will produce only one type of vehicle at a time, the purpose is to redesign the entire assembly plant in order to concentrate efforts and increase production.
- After selecting the vehicle that will be produced, the functional units required for assembly the vehicle are designed.
- After designing all the functional units, the assembly line is designed. Both (the assembly line and the control unit) will be a little bit simpler, because now they have less functional

units (remember that now only have the functional units required to assembly only a vehicle type).

- In order to increase the production will be replicated the assembly line as many times as possible and the number of replicates will be determined taking into account two aspects: in the first place, the speed with which the inputs can be placed at the beginning of the assembly line and secondly by the space available within the company.

5.1. FPGA architecture

An FPGA consists mainly of three types of components (see Figure 8): Configurable Logic Blocks (or Logic Array Block depending on vendor), reconfigurable interconnects and I/O blocks.

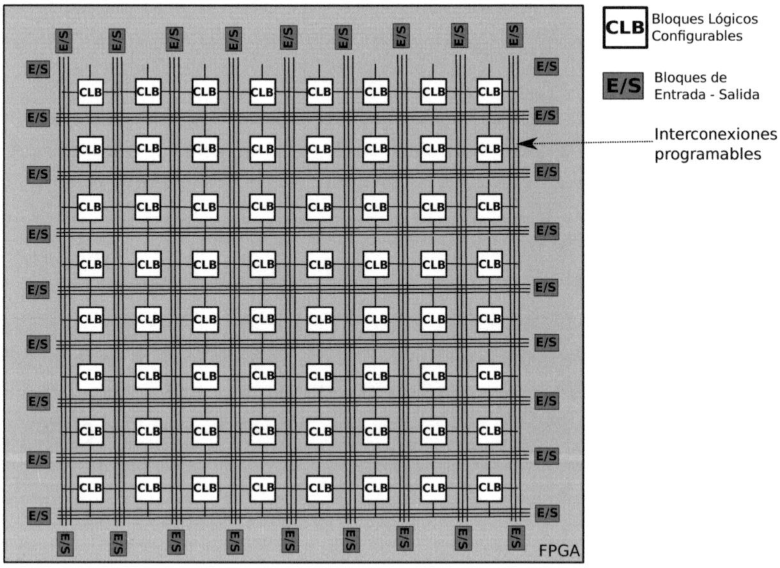

Figure 8. FPGA components.

In the analogy of the car assembly line, the Configurable Logic Blocks (CLBs) come to be the raw material for the construction of each one of the functional units required to assemble the car. In an FPGA the CLBs are used to build all the functional units required in an specific algorithm, these units can go from simple logical functions (and, or, not) to mathematical complex operations (algorithms, trigonometric functions, etc.). Therefore, a first step in the implementation of an algorithm over an FPGA is the construction of each one of the functional units required by the algorithm to be implemented (see figure 9). The design of these functional units is carried out by means of Hardware Description Languages (HDLs) like VHDL or Verilog.

The design of a functional unit using HDLs is not simply to write a code free of syntax errors, the process corresponds more to the design of a digital electronic circuit [13] which requires a good grounding about digital hardware. It must be remembered that this process consumes more time than an implementation on a CPU or a GPU.

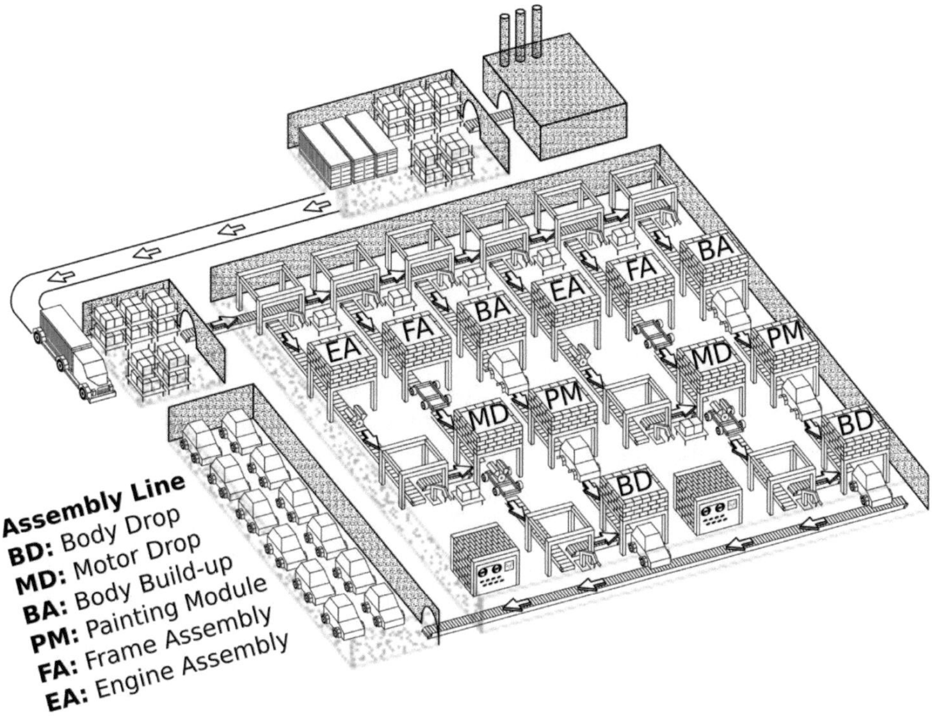

Figure 9. FPGA analogy.

Currently, providers of FPGAs (within which highlights Xilinx [42] and Altera [2]) offer predesigned modules that facilitate the design, in this way, commonly used blocks (as floating point units) should not be designed from scratch for each new application; these modules are offered as an HDL description (softcore modules) or physically implemented inside the FPGA (modules hardcore).

The reconfigurable interconnects, allows to interconnect the CLBs and the functional units each other, in the analogy of the car assembly line, the reconfigurable interconnects can be viewed as the conveyor belt where circulating the parts necessary for the assembly of the cars.

Finally, the I/O blocks allow to communicate the pins of the device and the internal logic of the FPGA, in the analogy of the car assembly line, the I/O blocks are the devices that place the parts at the beginning of the conveyor belt and also the devices that allow to bring out the cars of the factory. The blocks of input-output are responsible for controlling the flow of data between physical devices (extern) and the functional units in the interior of the FPGA. The different companies design these blocks in order to support different digital communication standards.

5.2. Algorithms that can be accelerated in an FPGA

Not all the applications can be benefited in the same way in an FPGA and due to the difficulty of implementing an application in these devices, it is advisable to do some previous estimates in order to determine the possibilities to speed up an specific algorithm.

Applications that require processing large amounts of data with little or no data dependency, [3] are ideal for implementing in the FPGAs, in addition it is required that these applications are limited by computing and not by data transfer, that is to say, the number of mathematical operations is greater than the number of read and write operations, [20]. In this regard, the SM have in favor that requires processing large amounts of data (in order of tens of Terabytes) and is required to perform billions of mathematical operations. However, migration also have a large number of read and write instructions which cause significant delays in the computational process.

Furthermore, the accuracy and the data format are other influential aspects on the performance of the applications over the FPGAs, with a lower data accuracy (least amount of bits to represent data), the performance of the application inside the FPGA increase; regarding the data format, the FPGAs get better performances with fixed point numbers than floating point numbers. The SM has been traditionally implemented using floating point numbers with single precision, [1, 21]. [14, 16] show that it is possible to implement some parts of the SM using fixed point instead of floating point and produce images with very similar patterns (visually identical), reducing computation times.

The FPGAs have a disadvantage in terms of the operation frequency, if they are compared with a CPU (10 times lower) or a GPGPU (5 times), for this reason, in order to accelerate an application inside an FPGA is required to perform at least 10 times more computational work per clock cycle than the performed in a CPU, [20]. In this regard, it is important that the algorithm has a great potential of parallelization. So in this aspect, it is helpful for the SM that the traces can be processed independently, due that the shots are made in different times, [4], which facilitates the parallelization of the process.

5.3. Implementation of the SM on FPGAs

Since 2004, it began to be implemented on the FPGAs, processes related with the construction of subsurface imaging. These implementations have made great contributions to the problems of processing speed and memory.

5.3.1. Breaking the power wall

As mentioned above, the operating frequency of traditional computing devices has been stalled for several years, due to problems related to the power dissipation. On the other hand, to mitigate the problems associated with the processing speed inside the FPGAs, it has been implemented modules that optimize the development of expensive mathematical operations[4], for example in [21] functional units were designed using the Cordic method, [3] to perform the square root, in [17] are used Lee approximations, [28] to perform trigonometric functions. Other research has addressed this problem from the perspective of the representation of numbers [17], the purpose is to change the single-precision (32-bit) floating-point format to fixed-point format (this is not possible, either for CPUs or GPUs) or a floating point format of less than 32 bits, in order to that the mathematical operations can be carried out in less time.

[3] The data dependency is one situation in which the instructions of a program require results of previous ones that have not yet been completed.

[4] The expensive mathematical operations are those that take more clock cycles to complete the process and consume more hardware resources; the expensive operations that are used in seismic processing are the square roots, logarithms and trigonometric functions, among others.

All these investigations have reported significant reductions in processing times of the SM sections when were compared with state of the art CPUs.

5.3.2. *Breaking the memory wall*

Computing systems based on FPGAs have both on-board memory[5] as on-chip memory [6], these two types of memory are faster than the external memories. Some research has made implementations in order to reduce the delays caused in the reading and writing operations [22], in their researches, have been designing special memory architectures that allow to optimize the use of different types of memory with FPGAs. The intent is that the majority of read and write operations must be performed at the speed of the on chip memory because this is the fastest, however, the challenge is to put all the data required in each instruction in on-chip memory because this is the one of the smaller capacity.

6. Wishlist

In spite of the great possibilities that have both FPGAs and GPGPUs to reduce the computation times of the seismic processes, their performance at this time is braking, because the rate at which seismic data are transmitted from the main memory to the computing device (FPGA or GPU) is not enough to maintain its computing potential 100% busy. The communication between the FPGAs or GPUs with the exterior can be carried out using large amount of output interfaces, currently one of the most used is Peripheral Component Interconnect Express (PCIe) port that transfer data between the computing device (FPGA or GPU) and a CPU at speeds of 2GB/sec, [14] and this transfer rate can not provide all the necessary data to keep the computing device busy.

Currently, at the Universidad Industrial de Santander (Colombia), a PhD project is being carried out that seeks to increase the transfer rate of seismic data between the main memory and the FPGA using data compression.

In addition, there are currently two approaches to try to reduce the design time on FPGAs (the main problem of this technology): The first strategy is to use compilers CtoHDL [11, 25, 35, 39, 43], these from a C code generates a description in a HDL, without having to manually perform the design; the second strategy is to use high level languages, these languages are the called Like-C as Mitrion-C [31] or SystemC [26]; despite their need, these languages have not yet achieved wide acceptance in the market, because its use still compromise the efficiency of the designs. The research regarding the compilation CtoHDL and high-level languages are active [29, 33, 34, 40, 44] and some results have been positive [7], but the possibility of having access to all the potential of the FPGAs from a high-level language is still in development.

On the other hand, it can be seen that the technological evolution of the GPUs is similar to the beginnings of the PCs evolution, where their progress was subject to Moore's law. It is therefore expected that in coming years new GPGPUs families continue to appear, increasingly so much that are going to raid in many areas of the HPC, but definitively it will come the time when this technology will stagnate for the same three barriers that stopped the

[5] This is the available memory on the board which contains the FPGA

[6] These are blocks of memory inside the FPGA

computers. When that time comes it will be expected that the FPGAs will be technological mature and can take the HPC baton during the next years.

At the end, we believe that the HPC future is going to be built of heterogeneous cluster, composed by these three technologies (CPUs, GPUs and FPGAs). These clusters will have an special operating system (O.S.) that will take advantage of each of these technologies, reducing the three barriers effect and getting the best performance in each application.

Acknowledgments

The authors wish to thank the Universidad Industrial de Santander (UIS), in particular the research group in Connectivity and Processing of Signals (CPS), head by Professor Jorge H. Ramon S. who has unconditionally been supporting our research work. Additionally we thank also to the Instituto Colombiano del Petroleo (ICP) for their constant support in recent years, especially thanks to Messrs William Mauricio Agudelo Zambrano and Andres Eduardo Calle Ochoa.

Author details

Sergio Abreo, Carlos Fajardo, William Salamanca and Ana Ramirez
Universidad Industrial de Santander, Colombia

7. References

[1] Abreo, S. & Ramirez, A. [2010]. Viabilidad de acelerar la migración sísmica 2D usando un procesador específico implementado sobre un FPGA The feasibility of speeding up 2D seismic migration using a specific processor on an FPGA, *Ingeniería e investigación* 30(1): 64–70.

[2] *Altera* [n.d.]. `http://www.altera.com/`. Reviewed: April 2012.

[3] Andraka, R. [1998]. A survey of cordic algorithms for fpga based computers, *Proceedings of the 1998 ACM/SIGDA sixth international symposium on Field programmable gate arrays*, FPGA '98, ACM, New York, NY, USA, pp. 191–200.
URL: *http://doi.acm.org/10.1145/275107.275139*

[4] Araya-polo, M., Cabezas, J., Hanzich, M., Pericas, M., Gelado, I., Shafiq, M., Morancho, E., Navarro, N., Valero, M. & Ayguade, E. [2011]. Assessing Accelerator-Based HPC Reverse Time Migration, *Electronic Design* 22(1): 147–162.

[5] Asanovic, K., Bodik, R., Catanzaro, B. C., Gebis, J. J., Husbands, P., Keutzer, K., Patterson, D. A., Plishker, W. L., Shalf, J., Williams, S. W. & Yelick, K. A. [2006]. The landscape of parallel computing research: A view from berkeley, *Technical Report UCB/EECS-2006-183*, EECS Department, University of California, Berkeley.
URL: *http://www.eecs.berkeley.edu/Pubs/TechRpts/2006/EECS-2006-183.html*

[6] Bednar, J. B. [2005]. A brief history of seismic migration, *Geophysics* 70(3): 3MJ–20MJ.

[7] Bier, J. & Eyre, J. [Second Quarter 2010]. BDTI Study Certifies High-Level Synthesis Flows for DSP-Centric FPGA Designs, *Xcell Journal Second* pp. 12–17.

[8] Bit-tech staff. [n.d.]. Intel sandy bridge review., `http://www.bit-tech.net/`. Reviewed: April 2012.

[9] Brodtkorb, A. R., Dyken, C., Hagen, T. R. & Hjelmervik, J. M. [2010]. State-of-the-art in heterogeneous computing, *Scientific Programming* 18: 1–33.

[10] Brodtkrorb, A. R. [2010]. *Scientific Computing on Heterogeneous Architectures*, Phd thesis, University of Oslo.

[11] C to Verilog: automating circuit design [n.d.]. `http://www.c-to-verilog.com/`. Revisado: Junio de 2011.

[12] Cabezas, J., Araya-Polo, M., Gelado, I., Navarro, N., Morancho, E. & Cela, J. M. [2009]. High-Performance Reverse Time Migration on GPU, *2009 International Conference of the Chilean Computer Science Society* pp. 77–86.

[13] Chu, P. [2006]. *RTL Hardware Design Using VHDL: Coding for Efficiency, Portability, and Scalability*, Wiley-IEEE Press.

[14] Clapp, R. G., Fu, H. & Lindtjorn, O. [2010]. Selecting the right hardware for reverse time migration, *The Leading Edge* 29(1): 48.
URL: *http://link.aip.org/link/LEEDFF/v29/i1/p48/s1&Agg=doi*

[15] Cleveland, C. [2006]. Mintrop, Ludger, *Encyclopedia of Earth* .

[16] Fu, H., Osborne, W., Clapp, R. G., Mencer, O. & Luk, W. [2009]. Accelerating seismic computations using customized number representations on fpgas, *EURASIP J. Embedded Syst.* 2009: 3:1–3:13.
URL: *http://dx.doi.org/10.1155/2009/382983*

[17] Fu, H., Osborne, W., Clapp, R. G. & Pell, O. [2008]. Accelerating Seismic Computations on FPGAs From the Perspective of Number Representations, *70th EAGE Conference & Exhibition* (June 2008): 9 – 12.

[18] Hagedoorn, J. [1954]. *A process of seismic reflection interpretation*, E.J. Brill.
URL: *http://books.google.es/books?id=U6FWAAAAMAAJ*

[19] Harris, D. M. & Harris, S. L. [2009]. *Digital Desing and Computer Architecture*, MK.

[20] Hauck Scott, D. A. [2008]. *Reconfigurable computing. The theory and practice of FPGA-BASED computing*, ELSEVIER - Morgan Kaufmann.

[21] He, C., Lu, M. & Sun, C. [2004]. Accelerating Seismic Migration Using FPGA-Based Coprocessor Platform, *12th Annual IEEE Symposium on Field-Programmable Custom Computing Machines* pp. 207–216.

[22] He, C., Zhao, W. & Lu, M. [2005]. Time domain numerical simulation for transient waves on reconfigurable coprocessor platform, *Proceedings of the 13th Annual IEEE Symposium on Field-Programmable Custom Computing Machines*, IEEE Computer Society, pp. 127–136.

[23] Hennessy, J. L. & Patterson, D. A. [2006]. *Computer Architecture A Quantitative Approach*, 4th edn, Morgan Kaufmann.

[24] IEEE Computer Society [n.d.]. Tribute to Seymour Cray, `http://www.computer.org/portal/web/awards/seymourbio`. Reviewed: April 2012.

[25] Impulse Accelerate Technologies [n.d.]. Impulse codeveloper c-to-fpga tools, `http://www.jacquardcomputing.com/roccc/`. Revisado: Agosto de 2011.

[26] Initiative, O. S. [n.d.]. Languaje SystemC, `http://www.systemc.org/home/`. Reviewed: April 2012.

[27] Intel [2011]. Intel core i7-3960x processor extreme edition, `http://en.wikipedia.org/wiki/FLOPS`. Reviewed: April 2012.

[28] Lee, D.-U., Abdul Gaffar, A., Mencer, O. & Luk, W. [2005]. Optimizing hardware function evaluation, *IEEE Trans. Comput.* 54: 1520–1531.
URL: *http://dl.acm.org/citation.cfm?id=1098521.1098595*

[29] Lindtjorn, O., Clapp, R. G., Pell, O. & Flynn, M. J. [2011]. Beyond traditional microprocessors for Geoscience High-Performance computing a pplications and Geoscience, *Ieee Micro* pp. 41–49.

[30] Mayne, W. H. [1962]. Common reflection point horizontal data stacking techniques, *Geophysics* 27(06): 927–938.
[31] Mitrionics [n.d.]. Languaje Mitrion-C, `http://www.mitrionics.com/`. Reviewed: April 2012.
[32] Moore, G. E. [1975]. Progress in digital integrated electronics, *Electron Devices Meeting, 1975 International*, Vol. 21, pp. 11–13.
[33] Necsulescu, P. I. [2011]. *Automatic Generation of Hardware for Custom Instructions*, PhD thesis, Ottawa, Canada.
[34] Necsulescu, P. I. & Groza, V. [2011]. Automatic Generation of VHDL Hardware Code from Data Flow Graphs, *6th IEEE International Symposium on Applied Computational Intelligence and Informatics* pp. 523–528.
[35] Nios II C-to-Hardware Acceleration Compilern [n.d.]. `http://www.altera.com/`. Revisado: Junio de 2011.
[36] Onifade, A. [2004]. History of the computer, *Conference of History of Electronics*.
[37] Owens, J. D., Houston, M., Luebke, D., Green, S., Stone, J. E. & Phillips, J. C. [2008]. Gpu computing, *Proceedings of the IEEE* 96(5): 879–899.
[38] Rieber, F. [1936]. Visual presentation of elastic wave patterns under various structural conditions, *Geophysics* 01(02): 196–218.
[39] Riverside Optimizing Compiler for Configurable Computing: Roccc 2.0 [n.d.]. `http://www.jacquardcomputing.com/roccc/`. Revisado: Junio de 2011.
[40] Sánchez Fernández, R. [2010]. *Compilación C a VHDL de códigos de bucles con reuso de datos*, Tesis, Universidad Politécnica de Cataluña.
[41] Stolt, R. H. [1978]. Migration by fourier transform, *Geophysics* (43): 23–48.
[42] *Xilinx* [n.d.a]. `http://www.xilinx.com/`. Reviewed: April 2012.
[43] Xilinx [n.d.b]. High-Level Synthesis: AutoESL, `http://www.xilinx.com/university/tools/autoesl/`. Reviewed: April 2012.
[44] Yankova, Y., Bertels, K., Vassiliadis, S., Meeuws, R. & Virginia, A. [2007]. Automated HDL Generation: Comparative Evaluation, *2007 IEEE International Symposium on Circuits and Systems* pp. 2750–2753.

Chapter 5

The Slug Flow Problem in Oil Industry and Pi Level Control

Airam Sausen, Paulo Sausen and Mauricio de Campos

Additional information is available at the end of the chapter

http://dx.doi.org/10.5772/50711

1. Introduction

The slug is a multiphase flow pattern that occurs in pipelines which connect the wells in seabed to production platforms in the surface in oil industry. It is characterized by irregular flows and surges from the accumulation of gas and liquid in any cross-section of a pipeline. In this work will be addressed the riser slugging, that combined or initiated by terrain slugging is the most serious case of instability in oil/water-dominated systems [5, 15, 21].

The cyclic behavior of the riser slugging, which is illustrated in Figure 1, can be divided into four phases: (i) Formation: gravity causes the liquid to accumulate in the low point in pipeline-riser system and the gas and liquid velocity is low enough to enable for this accumulation; (ii) Production: the liquid blocks the gas flow and a continuous liquid slug is produced in the riser, as long as the hydrostatic head of the liquid in the riser increases faster than the pressure drop over the pipeline-riser system, the slug will continue to grow; (iii) Blowout: when the pressure drop over the riser overcomes the hydrostatic head of the liquid in the riser the slug will be pushed out of the system; (iv) liquid fall back: after the majority of the liquid and the gas has left the riser the velocity of the gas is no longer high enough to drag the liquid upwards, the liquid will start flowing back down the riser and the accumulation of liquid starts over again [15].

The slug flow causes undesired consequences in the whole oil production such as: periods without liquid or gas production into the separator followed by very high liquid and gas rates when the liquid slug is being produced, emergency shutdown of the platform due to the high level of liquid in the separators, floods, corrosion and damages to the equipments of the process, high costs with maintenance. One or all these problems cause significant losses in oil industry. The main one has been of economic order, due to reduction in oil production capacity [6, 8, 16–20].

Currently, control strategies are considered as a promising solution to handle the slug flow [4, 5, 7, 10, 15, 20]. An alternative to the implementation of control strategies is to make use of a mathematical model that represents the dynamic of slug flow in pipeline-separator system.

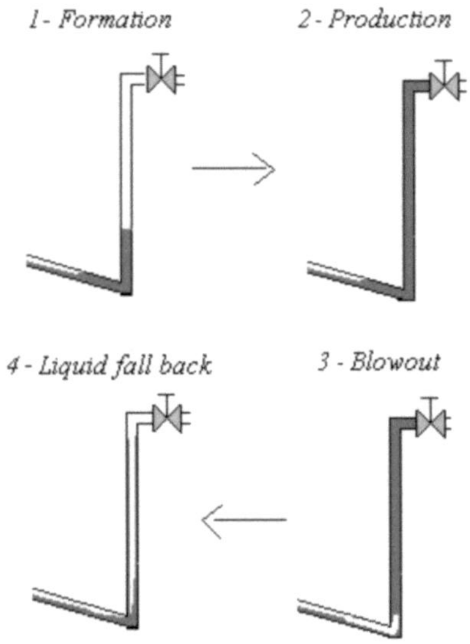

Figure 1. Illustration of a slug cycle.

In this chapter has been used the dynamic model for a pipeline-separator system under the slug flow, with 5 (five) Ordinary Differential Equations (ODEs) coupled, nonlinear, 6 (six) tuning parameters and more than 40 (forty) internal, geometric and transport equations [10, 13], denominated Sausen's model.

To carry out the simulation and implementation of control strategies in the Sausen's model, first it is necessary to calculate its tuning parameters. For this procedure, are used data from a case study performed by [18] in the OLGA commercial multiphase simulator widely used in the oil industry. Next it is important to check how the main variables of the model change their behavior considering a change in the model's tuning parameters. This testing, called sensitivity analysis, is an important tool to the building of the mathematical models, moreover, it provides a better understanding of the dynamic behavior of the system, for later implementation of control strategies.

In this context, from the sensitivity analysis, the Sausen's model has been an appropriate environment for application of the different feedback control strategies in the problem of the slug in oil industries through simulations. The model enables such strategies can be applied in consequence the slug, that is in the oil or gas output valve separator, as well as in their causes, in the top riser valve, or yet in the integrated system, in other words, in more than one valve simultaneously.

Therefore, as part of control strategies that can be used to avoid or minimize the slug flow, this chapter presents the application of the error-squared level control strategy Proportional Integral (PI) in the methodology by bands [10], whose purpose damping of the load flow rate oscillatory that occur in production's separators. This strategy is compared with the level controls strategy PI conventional [1], widely used in industrial processes; and with the level control strategy PI also in the methodology by bands.

The remainder of this chapter is organized as following. Section 2 presents the equation of the Sausen's model for a pipeline-separator system. Section 3 shows the simulation results of the Sausen's model. Section 4 presents the control strategies used to avoid or minimize the slug flow. Section 5 shows the simulation results and analysis of the control strategies applied the Sausen's model. And finally, in Section 6, are discussed the conclusions and future research directions.

2. The dynamic model

2.1. Introduction

This section presents a mathematical model for the pipeline-separator system, illustrated in Figure 2 with biphase flow (gas-liquid). The model is the result of coupling the simplified dynamic model of Storkaas [15, 17, 20] with the model for a biphase horizontal cylindrical separator [22]. The new model has been called of Sausen's model.

The following are shown the modelling assumptions, the model equations, how the distribution of liquid and gas inside the pipeline for the separator occurs. Finally, are presented the simulation results for this model considering two settings for simulations: (i) the opening valve Z in top of the riser $z = 20\%$ (flow steady); and (ii) the opening valve Z in top of the riser $z = 50\%$ (slug flow).

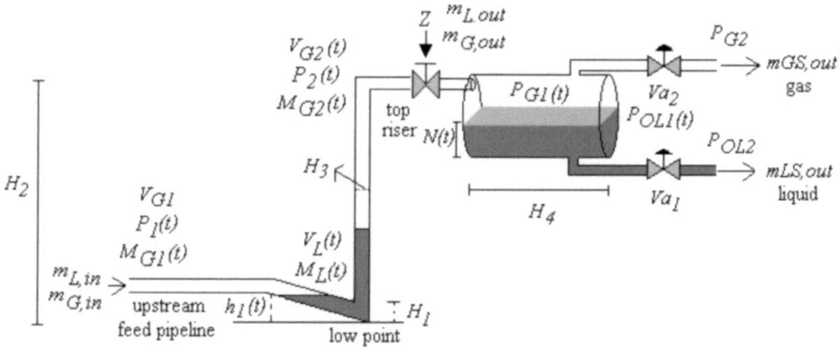

Figure 2. Illustration of the pipeline-separator system with the slug formation.

2.2. Model assumption

The Sausen's model assumptions are presented as follow.

A1: Liquid dynamics in the upstream feed section of the pipeline have been neglected, that is, the liquid velocity in this section is constant.

A2: Follows from assumption A1 that the gas volume is constant in the upstream feed section pipeline and that the volume variations due to liquid level $h_1(t)$ at the low point are neglected.

A3: Only one dynamical state $M_L(t)$ for holdup liquid in the riser section. This state includes both the liquid in the riser and at the low point section (with level $h_1(t)$).

A4: Two dynamical states for holdup gas (M_{G1} and $M_{G2}(t)$) occupying the volumes V_{G1} and $V_{G2}(t)$, respectively. The gas volumes are related to each other by a pressure-flow relationship at the low point.

A5: Simplified valve equation for gas and liquid mixture leaving the system at the top of the riser.

A6: Stationary pressure balance over the riser (between pressures $P_1(t)$ and $P_2(t)$).

A7: There is not chemical reaction between the fluids (gas-liquid) in pipeline.

A8: Each one of the fluid consists of a single component in the separator.

A9: The portion of liquid mixed with the gas in the entrance of the separator is neglected.

A10: Simplified valve equation for the gas and the liquid leaving the separator.

A11: The liquid is incompressible.

A12: The temperature is constant.

A13: The gas has ideal behavior.

2.3. Model equations

The Sausen's model is composed of 5 (five) ODEs that are based on the mass conservation equations. The equations (1)-(3) represent the dynamics of the pipeline system and the equations (4)-(5) represent the dynamics of the separator:

$$\dot{M}_L(t) = m_{L,in} - m_{L,out}(t) \tag{1}$$

$$\dot{M}_{G1}(t) = m_{G,in} - m_{Gint}(t) \tag{2}$$

$$\dot{M}_{G2}(t) = m_{G1}(t) - m_{G,out}(t) \tag{3}$$

$$\dot{N}(t) = \frac{\sqrt{r_S^2 - (r_s - N(t))^2}}{2H_4\rho_L N(t)\,[3r_s - 2N(t)]}\left[m_{L,out}(t) - m_{LS,out}(t)\right] \tag{4}$$

$$\dot{P}_{G1}(t) = \frac{\{\rho_L\Phi[m_{G,out}(t) - m_{GS,out}(t)] + P_{G1}(t)\,[m_{L,out}(t) - m_{LS,out}(t)]\}}{\rho_L[V_S - V_{LS}(t)]} \tag{5}$$

where: $M_L(t)$ is the liquid mass at low point in the pipeline, (kg); $M_{G1}(t)$ is the gas mass in the upstream feed section of pipeline, (kg); $M_{G2}(t)$ is the gas mass at the top of the riser, (kg); $N(t)$ is the liquid level in the separator, (m); $P_{G1}(t)$ is the gas pressure in the separator, (N/m^2); and the $\dot{M}_L(t)$, $\dot{M}_{G1}(t)$, $\dot{M}_{G2}(t)$, $\dot{N}(t)$, $\dot{P}_{G1}(t)$ are their respective derivatives in relation to time; $m_{L,in}$ is the liquid mass flowrate that enters the upstream feed section of the pipeline, (kg/s); $m_{G,in}$ is the gas mass flowrate that enters in the upstream feed section of the pipeline, (kg/s); $m_{L,out}(t)$ is the liquid mass flowrate leaving through the valve at the top of the riser enters the separator, (kg/s); $m_{G,out}(t)$ is the gas mass flowrate leaving through the valve at the top of the riser enters the separator, (kg/s); $m_{Gint}(t)$ is the internal gas mass flowrate, (kg/s); $m_{LS,out}(t)$ is the liquid mass flowrate that leaves the separator through the valve Va_1, (kg/s); $m_{GS,out}(t)$ is the gas mass flowrate that leaves the separator through the valve Va_2, (kg/s); r_s is the separator ray, (m); H_4 is the separator length, (m); ρ_L is the liquid density, (kg/m^3); V_S is the separator volume, (m^3); $V_{LS}(t)$ is the liquid volume in the separator, (m^3); $\Phi = \frac{RT}{M_G}$ is a constant; R is the ideal gas constant ($8314\ \frac{J}{K.kmol}$); T is the temperature, (K); M_G is the gas molecular weight, $(kg/kmol)$.

The stationary pressure balance over the riser is assumed to be given by

$$P_1(t) - P_2(t) = g\bar{\rho}(t)H_2 - \rho_L g h_1(t)$$

where: $P_1(t)$ is the gas pressure in the upstream feed section of the pipeline, (N/m^2); $P_2(t)$ is the gas pressure at the top of the riser, (N/m^2); g is the gravity $(9.81m/s^2)$; $\bar{\rho}(t)$ is the average mixture density in the riser, (kg/m^3); H_2 is the riser height, (m); $h_1(t)$ is the liquid level at the decline, (m).

A simplified valve equation is used to describe the flow through the Z valve at the top of the riser that is given by

$$m_{mix,out}(t) = zK_1\sqrt{\rho_T(t)(P_2(t) - P_{G1}(t))} \tag{6}$$

where: z is the valve position $(0 - 100\%)$; K_1 is the valve constant and a tuning parameter; $\rho_T(t)$ is the density upstream valve, (kg/m^3); $P_{G1}(t)$ is the gas pressure into the separator, (N/m^2). It is possible to observe that the coupling between the pipeline and the separator occurs through a pressure relationship, in other words, the gas pressure into the separator $P_{G1}(t)$ is the pressure before the Z valve at the top of the riser, according to equation (6).

Considering the result that has been shown in equation (6), it is also possible to obtain the liquid mass flowrate given by

$$m_{L,out}(t) = \alpha_L^m(t)m_{mix,out}(t)$$

and the gas mass flowrate given by

$$m_{G,out}(t) = [1 - \alpha_L^m(t)]m_{mix,out}(t)$$

that leave through the Z valve at the top of the riser, where $\alpha_L^m(t)$ is the liquid fraction upstream valve.

The liquid mass flowrate that leaves the separator is represented by the Va_1 valve equation given by

$$m_{LS,out}(t) = z_L K_4\sqrt{\rho_L[P_{G1}(t) + g\rho_L N(t) - P_{OL2}]} \tag{7}$$

where: z_L is the liquid valve opening $(0 - 100\%)$; K_4 is the valve constant and a tuning parameter; P_{OL2} is the downstream pressure after the V_{a1} valve, (N/m^2).

The gas mass flowrate that leaves the separator is represented by the Va_2 valve equation given by

$$m_{GS,out}(t) = z_G K_5\sqrt{\rho_G(t)[P_{G1}(t) - P_{G2}]} \tag{8}$$

where: z_G is the gas valve position $(0 - 100\%)$; K_5 is the valve constant and a tuning parameter; $\rho_G(t)$ is the gas density, (kg/m^3); P_{G2} is the downstream pressure after the V_{a2} valve, (N/m^2).

The boundary condition at the inlet (inflow $m_{L,in}$ and $m_{G,in}$) can either be constant or dependent on the pressure. In this work they are constant and have been considered disturbances of the process. The most critical section of the model is the phase distribution and phase velocities of the fluids in the pipeline-riser system. The gas velocity is based on an assumption of purely frictional pressure drop over the low point and the liquid distribution is based on an entrainment model. Finally, the internal, geometric and transport equations for the pipeline system are found in [15, 17, 20].

2.4. Displacement of the gas flow

The displacement of gas in the pipeline system occurs through a relationship between the gas mass flow and the variation of the pressure inside the pipeline. The acceleration has been neglected for the gas phase, so that it is the difference of the pressure that makes the fluids outflow pipeline above. Its equation is given by

$$\Delta P(t) = P_1(t) - [P_2(t) + g\rho_L \alpha_L(t) H_2]$$

where: $\alpha_L(t)$ is the average liquid fraction in riser.

It is considered that there are two situations in the riser: (i) $h_1(t) > H_1$, in this case the liquid is blocking the low point and the internal gas mass flowrate $m_{Gint}(t)$ is zero; (ii) $h_1(t) < H_1$, in this case the liquid is not blocking the low point, so the gas will flow from V_{G1} to $V_{G2}(t)$ with a internal gas mass flowrate $m_{Gint}(t) \neq 0$, where V_{G1} is the gas volume in upstream feed section of the pipeline, (m^3) and V_{G2} is the gas volume at the top of the riser, (m^3).

From physical insight, the two most important parameters determining the gas flowrate are the pressure drop over the low point and the free area given by the relative liquid level

$$\xi(t) = (H_1 - h_1(t))/H_1$$

at the low point. This suggests that the gas transport could be described by a valve equation, where the pressure drop is driving the gas through a valve with opening $\xi(t)$. Based on trial and error, the following valve equation has been proposed

$$m_{G1}(t) = K_2 f(h_1(t)) \sqrt{\rho_{G1}(t)[P_1(t) - P_2(t) - g\rho_L \alpha_L(t) H_2]} \tag{9}$$

where: K_2 is the valve constant and a tuning parameter; $f(h_1(t)) = \hat{A}(t)\xi(t)$ e $\hat{A}(t)$ is the cross-section area at the low point, (m^2); $h_1(t)$ is the liquid level upstream in the decline, (m); H_1 is the critical liquid level, (m); $\rho_{G1}(t)$ is the gas density in the volume 1, (kg/m^3). The internal gas mass flowrate from the volume V_{G1} to volume $V_{G2}(t)$ is given by

$$m_{Gint}(t) = v_{G1}(t)\rho_{G1}(t)\hat{A}(t) \tag{10}$$

where: $v_{G1}(t)$ is the gas velocity at the low point, m/s. Therefore, substituting equation (10) into equation (9), it has been found that the gas velocity is

$$v_{G1}(t) = \begin{cases} K_2\xi(t)\sqrt{\frac{P_1(t) - P_2(t) - g\rho_L\alpha_L(t)H_2}{\rho_{G1}(t)}} & \forall h_1(t) < H_1, \\ 0 & \forall h_1(t) \geq H_1. \end{cases} \tag{11}$$

2.5. Entrainment equation

The distribution of liquid occurs through an entrainment equation. It is considered that the gas pushes the liquid riser upward, then the volume fraction of liquid ($\alpha_{LT}(t)$) that is leaving through the Z valve at the top of the riser is modelled.

The volume fraction of liquid will lie between two extremes: (i) when the liquid blocks the gas flow ($v_{G1} = 0$), there is no gas flowing through the riser and $\alpha_{LT}(t) = \alpha^*_{LT}(t)$, in most cases there will be only gas leaving the riser, so $\alpha^*_{LT}(t) = 0$, however, eventually the entering liquid may cause the liquid to fill up the riser and $\alpha^*_{LT}(t)$ will exceed zero; (ii) when the gas velocity is very high there will be no slip between the phases, so $\alpha_{LT}(t) = \alpha_L(t)$, where $\alpha_L(t)$ is the average liquid fraction in the riser.

The transition between these two extremes should be smooth and occurs as follows: when the liquid blocks the low point of the riser, the liquid fraction on top is $\alpha^*_{LT}(t) = 0$, so the amount of liquid in the riser goes on increases until $\alpha^*_{LT}(t) > 0$. At this moment the gas pressure and the gas velocity in the feed upstream section of the pipeline is very high, then the entrainment occurs. This transition depends on a parameter $q(t)$. The entrainment equation is given by

$$\alpha_{LT}(t) = \alpha^*_{LT}(t) + \frac{q^n(t)}{1 + q^n(t)}(\alpha_L(t) - \alpha^*_{LT}(t)) \qquad (12)$$

where

$$q(t) = \frac{K_3 \rho_{G1}(t) v^2_{G1}(t)}{\rho_L - \rho_{G1}(t)}$$

and K_3 and n are tuning parameters of the model. The details of the modelling of the equation (12) are found in Storkaas [15].

3. Simulation and analysis results of the Sausen's model

In this section are presented the simulation results of the Sausen's model for a pipeline-separator system. Initially the tuning parameters are calculated: K_1 in Z valve equation (6), K_2 in gas velocity equation (11), K_3 and n in entrainment equation (12), K_4 in Va_1 liquid valve equation (7), and K_5 in Va_2 gas valve equation (8). The calculation of these tuning parameters depends on the available data from a real system or an experimental loop, but a complete set of data is not found in the literature and is not provided by oil industries.

Therefore, to calculate the tuning parameters of the dynamic model are used the case study data carried out by Storkaas [15] through the multiphase commercial simulator OLGA [2] that accurately represents the pipeline system under slug flow [15] and the data of separator dimensioned from a tank of literature [10]. In this case study the transition of the steady flow to a slug flow occurs in the valve opening $z = 13\%$ (i.e., $z_{crit} = 13\%$). Table 1 presents the data for the simulation of the dynamic model and Table 2 presents the values of the tuning parameters of the dynamic model.

Now are presented the simulation results considering the Z valve opening $z = 12\%$. Figure 3 shows the varying pressures $P_1(t)$ in the upstream feed section and $P_2(t)$ at the top of the riser. Figure 4 shows the dynamics of the liquid mass flowrate (up-left) and the dynamics of the gas mass flowrate (down-left) that are entering the separator, and the dynamics of the liquid mass flowrate (up-right) and the dynamics of the gas mass flowrate (down-right) that are leaving the separator. Figure 5 shows the dynamics of the liquid level (left) and of the gas pressure (right) in the separator. It is possible to observe in all these simulation results that the varying pressures induce oscillations, but because the valve position is less than z_{crit}, these oscillations eventually die out characterizing the steady flow in pipeline-separator system.

Symbol/Value	Description	SI
$m_{L,in} = 8.64$	Liquid mass flowrate into system	kg/s
$m_{G,in} = 0.362$	Gas mass flowrate into system	kg/s
$P_1(t) = 71.7 \times 10^5$	Gas pressure in the upstream feed section of the pipeline	N/m^2
$P_2(t) = 53.5 \times 10^5$	Gas pressure at the top of the riser	N/m^2
$r = 0,06$	Pipeline ray	m
$H_2 = 300$	Height of riser	m
$L_1 = 4300$	Length of horizontal pipeline	m
$L_3 = 100$	Length of horizontal top section	m
$H_4 = 4.5$	Length of separator	m
$D_s = 1.5$	Diameter of separator	m
$N_t = 0.75$	Liquid level	m
$P_{G1} = 50 \times 10^5$	Pressure after Z valve at the top of the riser	N/m^2
$P_{OL2} = 49 \times 10^5$	Pressure after Va_1 liquid valve of separator	N/m^2
$P_{GL2} = 49 \times 10^5$	Pressure after Va_2 gas valve of separator	N/m^2

Table 1. Data for simulation dynamic model.

φ	K_1	K_2	K_3	K_4	K_5
2.55	0.005	0.8619	1.2039	0.002	0.0003

Table 2. Model tuning parameters.

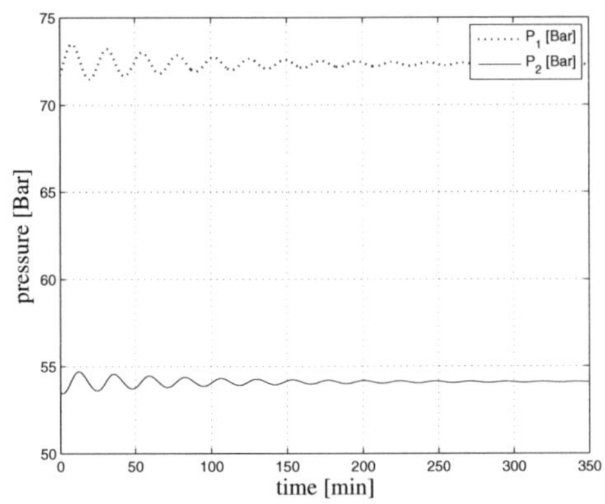

Figure 3. Varying pressures in pipeline system with $z = 12\%$ (steady flow).

In the following section we are presenting the simulation results considering the Z valve opening $z - 50\%$. Figure 6 shows the varying pressures throughout the pipeline system. Figure 7 presents the dynamics of the liquid mass flowrate (up-left) and the dynamics of the gas mass flowrate (down-left) that are entering the separator with peak mass flowrate of the 14 kg/s for the liquid and 2 kg/s for the gas, and the dynamics of the liquid mass flowrate (up-right) and the dynamics of the gas mass flowrate (down-right) that are leaving the separator. Figure 8 shows the dynamics of the liquid level (left) and of the gas pressure (right) in the separator. Finally, it has been observed that in all these simulation results the varying pressures induce periodical oscillations, characterizing the slug flow that happens in the pipeline-separator system. It has also been shown that the slug flow happens in intervals of 12 minutes.

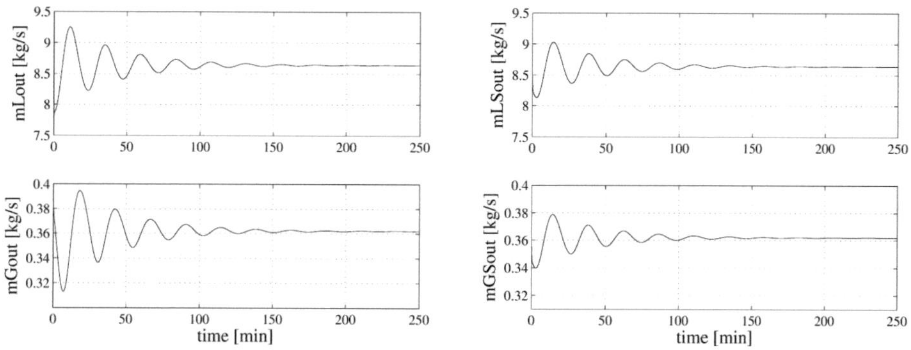

Figure 4. Input liquid (up-left) and input gas (down-left) mass flowrate in the separator and output liquid (up-right) and output gas (down-right) mass flowrate in the separator with $z = 12\%$ (steady flow).

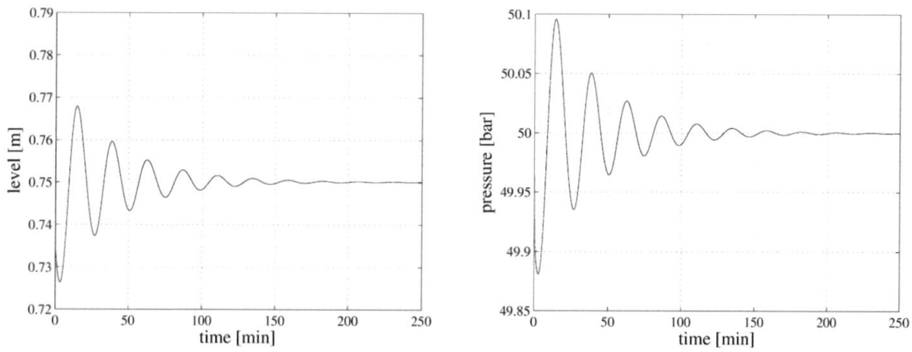

Figure 5. Liquid level (left) and gas pressure (right) in the separator with $z = 12\%$ (steady flow).

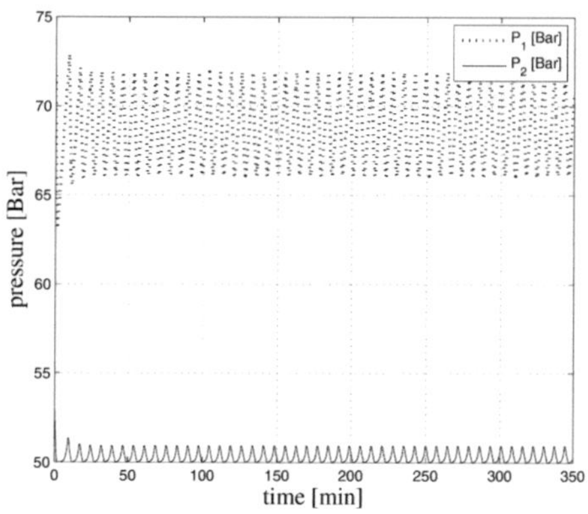

Figure 6. Varying pressures in the pipeline system with $z = 50\%$ (slug flow).

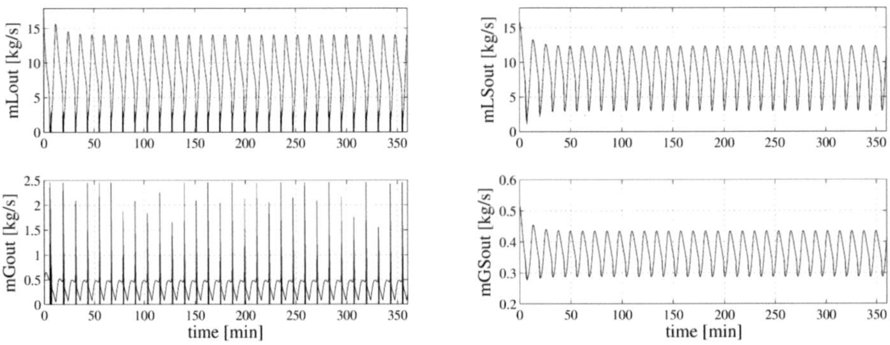

Figure 7. Input liquid (up-left) and gas (down-left) mass flowrate in the separator and output liquid (up-right) and gas (below-right) mass flowrate in the separator with $z = 50\%$ (slug flow).

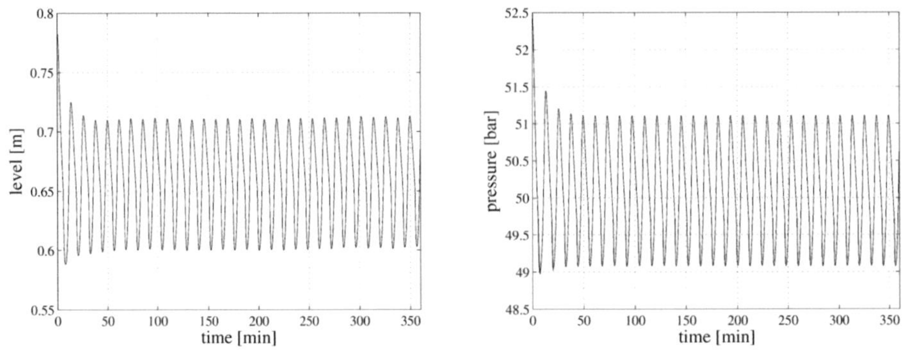

Figure 8. Liquid level (left) and gas pressure (right) in the separator with $z = 50\%$ (slug flow).

4. Control strategies

Controller PI actuating in the oil output valve in the oil industry is the traditional method used to control the liquid level in production separators. If the controller is tuned to maintain a constant liquid level, the inflow variations will be transmitted to the separator output, in this case, causing instability in the downstream equipments. An ideal liquid level controller will let the level to vary in a permitted range (i.e., band) in order to make the outlet flow more smooth; this response specification cannot be reached by PI controller conventional for slug flow regime. Nunes [9] defined a denominated level control methodology by bands, which promotes level oscillations within certain limits, i.e, the level can vary between the maximum and the minimum of a band, as Figure 9, so that the output flowrate is close to the average value of the input flowrate. This strategy does not use flow measurements and can be applied to any production separator.

In the band control when the level is within the band, it is used the moving average of the control action of a slow PI controller, because reducing the capacity of performance of the controller gives a greater fluctuation in the liquid level within the separator. The moving average is calculated in a time interval, this interval should be greater than the period T of

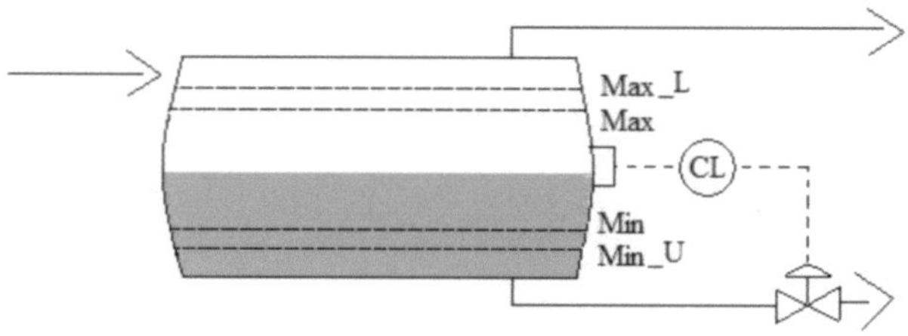

Figure 9. Band control diagram of *Nunes*.

the slug flow. When the band limits are exceeded, the control action in moving average of the slow PI controller is switched to the PI controller of the fast action for a time, whose objective is return to the liquid level for within the band, if so, the action of the control again will be the moving average. To avoid abrupt changes in action control for switching between modes of operation within the band and outside the band, it is suggested to use the average between the actions of PI controller of the fast action and in moving average.

Therefore, this paper performs the application in the Sausen's model of the level control strategy PI considering 3 (three) methodologies: (1) level control strategy PI conventional, the level shall remain fixed at setpoint; (2) level control strategy PI in the methodology by bands; (3) error-squared level control strategy PI in the methodology by bands.

The error-squared controller [14] is a continuous nonlinear function whose gain increases with the error. Its gain is computed as

$$k_c(t) = k_1 + k_{2NL}|e(t)|$$

where k_1 is a linear part, k_{2NL} is a nonlinear one and $e(t)$ is the tracking error. If $k_{2NL} = 0$ the controller is linear, but with $k_{2NL} > 0$ the function becomes squared-law.

In literature the error-squared controller is suggested to be used in liquid level control in production separators under load inflow variations. From the application of the error-squared controller, in liquid level control process in vessels, it is observed that small deviations from the setpoint resulted in very little change to the valve leaving the output flow almost unchanged. On the other hand large deviations are opposed by much stronger control action due to the larger error and the law of the error-squared, thereby preventing the level from rising too high in the vessel. The error-squared controller has the benefit of resulting in more steady downstream flow rate under normal operation with improved response when compared to the level control strategy conventional [12].

For implementation of the controllers it is used the algorithm control PI [1] in speed form, whose equation is given by

$$\Delta u(t) = k_c \Delta e(t) + k_c \frac{1}{T_i} T_a e(t) \tag{13}$$

where $\Delta u(t)$ is the variation of the control action; k_c is the gain controller; $\Delta e(t)$ is the variation of the tracking error; T_a is the sampling period of the controller; $e(t)$ is the tracking error. It is considered that the valve dynamics, i.e., the time for its opening reach the value of the control action is short, so this implies that the valve opening is the control action itself.

5. Simulation and analysis results of the control strategies

This section presents the simulation results of the control strategies using the computational tool Matlab. To implement the control by bands it is used a separator with length 4.5 m and diameter $1.5m$ following the standards used by [9]. The setpoint for the controller is $0.75mn$(i.e., separator half), the band is $0.2m$, where the liquid level maximum permitted is $0.95m$ and the minimum is $0.55m$. The bands were defined to follow the works of [3, 9]

Initially, for the first simulation, it is considered the Z valve opening at the top of the riser in $z = 20\%$ (slug flow). To simulate the level control strategy PI conventional the values used for the controller gain k_c and the integral time T_i are 10 and 1380s respectively, according to the heuristic method to tune level controllers proposed by Campus et al. (2006). In level control strategy PI in the methodology by bands the level can float freely within the band limits in separator. In this case, the controller PI with slow acting (i.e., within the band) uses controller gain $k_c = 0.001$ and integral time $T_i = 100000s$, and the PI controller with fast acting (i.e., out the band) uses $k_c = 0.15$ and $T_i = 1000s$. In error-squared control strategy PI in the methodology by bands, the gain linear and nonlinear of the controller are computed to following the methodology present in [11] based on Lyapunov stability theory. In this case, the error-squared level PI controller with slow acting (i.e., within the band) uses $k_c = 0.001$, $k_{2NL} = 0.000004$ and $T_i = 100000s$, and the PI controller with fast acting (i.e., out the band) uses $k_c = 0.15$, $k_{2NL} = 0.03$ and $T_i = 1000s$. The period for calculating the moving average of the PI controllers by band was $T_i = 1000s$.

Figure 10 (a) presents the liquid level variations $N(t)$ considering level controller strategy PI conventional (dashed line) and level control strategy PI in the methodology by bands (solid line) in the separator, and the Figure 10 (b) presents liquid level variations $N(t)$ considering level control strategy PI conventional (dashed line) and error-squared level control strategy PI in methodology by bands (solid line). Figures 11 (a) and (b) shown the liquid output flow rate variations $m_{LS,out}(t)$ of the separator that corresponding to controls of the presented in Figures 10 (a) and (b).

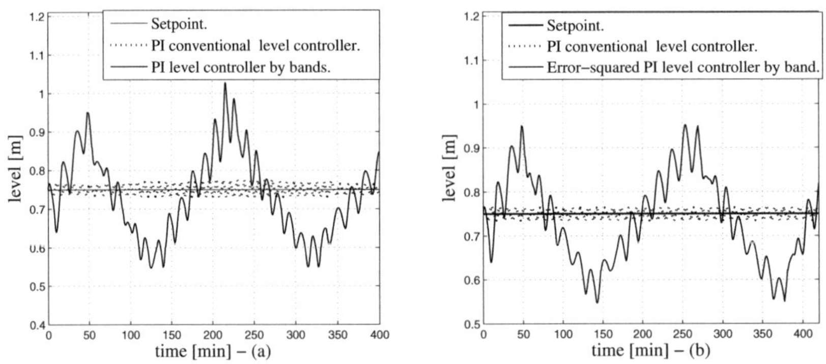

Figure 10. Liquid level variations $N(t)$, (a) level control strategy PI conventional (dashed line) and level control strategy PI by band (solid line), (b) level control strategy PI conventional (dashed line) and error-squared level control strategy PI by band (solid line), $z = 20\%$.

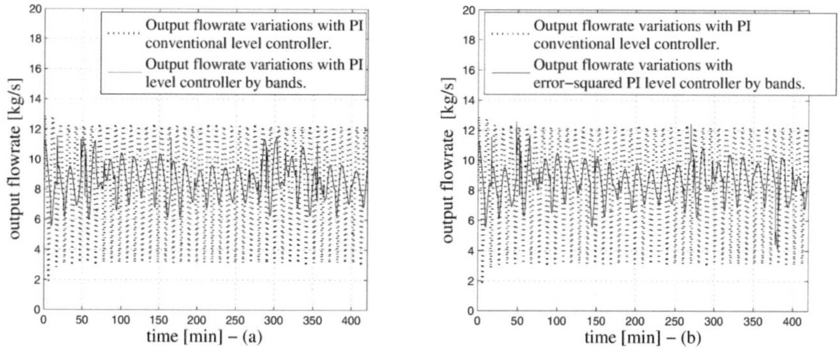

Figure 11. Liquid output flow rate variations $m_{LS,out}(t)$, (a) level control strategy PI conventional (dashed line) and level control strategy PI by band (solid line), (b) level control strategy PI conventional (dashed line) and error-squared level control strategy PI by band (solid line), $z = 20\%$.

The following are presented simulation results for valve opening at the top of the riser, i.e., $z = 20\%$, $z = 25\%$, $z = 30\%$ and $z = 35\%$ (slug flow). Figure 12 (a) presents the liquid level variations $N(t)$ considering level control strategy PI conventional (dashed line) and level control strategy PI in the methodology by bands (solid line) in separator, and the Figure 9 (b) presents liquid level variations $N(t)$ considering level control strategy PI conventional (dashed line) and error-squared level control strategy PI in the methodology by bands (solid line). Figures 10 (a) and (b) shown the liquid output flow rate variations of the separator that corresponding to controls of the level presented in Figure 9 (a) and (b).

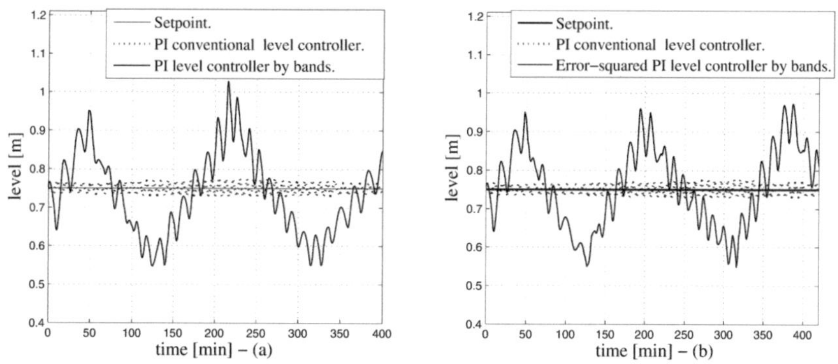

Figure 12. Liquid level variations $N(t)$, (a) level control strategy PI conventional (dashed line) and level control strategy PI by band (solid line), (b) level control strategy PI conventional (dashed line) and error-squared level control strategy PI by band (solid line), $z = 20\%$.

Comparing the simulation results between the level control strategy PI and error-squared level control strategy PI both in the methodology by bands, it is observed that the second controller (Figures 10 and 12 (b)) has respected strongly the defined bands, i.e., in $0.95m$ (higher band) and in $0.55m$ (lower band), because it has the more hard control action than the first controller (Figure 10 and 12 (a)). However, when the liquid level reached the band limits for the error-squared level control strategy PI, at this time, the liquid output flow rate has a little more oscillatory flows than the ones found for the level controller PI by bands, but

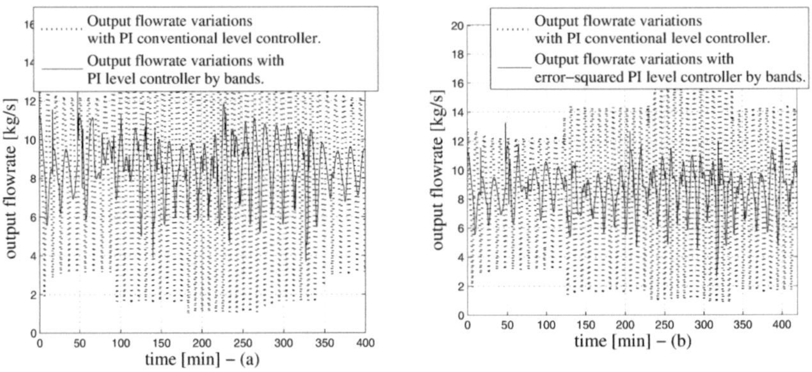

Figure 13. Liquid output flow rate variations $m_{LS,out}(t)$, (a) level control strategy PI conventional (dashed line) and level control strategy PI by band (solid line), (b) level control strategy PI conventional (dashed line) and error-squared level control strategy PI by band (solid line), $z = 20\%$.

this difference is minimal, according to Figures 11 (a) and (b), Figures 13 (a) and (b). For both controllers simulation results of the liquid output flow rate are better than the results obtained with the level control strategy PI conventional. Considering the liquid output flow rate when the level is within the band, both processes (i.e., level control strategy PI and error-squared level control strategy PI both in the methodology by band) have similar trends.

6. Conclusion

In this chapter with objective of reducing the export oscillatory flow rate caused by slug flow, three methodologies of the level controls were implemented (1) level control strategy PI conventional; (2) level control strategy PI in the methodology by bands; (3) error-squared level control strategy PI in the methodology by bands.

The simulation results showed that the error-squared level control PI strategy in the methodology for bands presented the better results when compared with the level control strategy PI conventional, because reduced flow fluctuations caused by slug flow; and with the level control strategy PI in the methodology by bands, it probably happened because the first has highly respected the defined bands.

As suggestions for future work new control strategies can be implemented in integrated system, i.e., more than one valve simultaneously. Considering the mathematical modeling of the process, it was necessary to investigate a mathematical model with fewer parameters, along with the construction of an experimental platform, since the data of a real process is difficult to obtain and are not provided by oil industry.

Author details

Airam Sausen, Paulo Sausen, Mauricio de Campos
Master's Program in Mathematical Modeling (MMM), Group of Industrial Automation and Control, Regional University of Northwestern Rio Grande do Sul State (UNIJUÍ), Ijuí, Brazil.

7. References

[1] Astrom, K. J. & Hagglund, T. [1995]. *PID Controllers: Theory, Design, and Tuning*, ISA, New York.

[2] Bendiksen, K., Malnes, D., Moe, R. & Nuland, S. [1991]. The dynamic two-fluid model olga: theory and application, *SPE Production Engineering*, pp. 171–180.

[3] de Campos, M. C. M., Costa, L. A., Torres, A. E. & Schmidt, D. C. [2008]. Advanced control levels of separators production platforms, *1 CICAP Congresso de Instrumentação, Controle e Automação da Petrobrás (I CICAP)*, Rio de Janeiro. in Portuguese.

[4] Friedman, Y. Z. [1994]. Tuning of averaging level controller, *Hydrocarbon Processing Journal* .

[5] Godhavn, M. J., Mehrdad, F. P. & Fuchs, P. [2005]. New slug control strategies, tuning rules and experimental results, *Journal of Process Control* 15: 547–577.

[6] Havre, K. & Dalsmo, M. [2002]. Active feedback control as a solution to severe slugging, *SPE Production and Facilities, SPE 79252* pp. 138–148.

[7] Havre, K., Stornes, K. & Stray, H. [2000]. Taming slug flow in pipelines, *ABB review*, number 4, pp. 55–63.

[8] Havre, K. & Stray, H. [1999]. Stabilization of terrain induced slug flow in multiphase pipelines, *Servomotet*, Trondheim.

[9] Nunes, G. C. [2004]. Bands control: basic concepts and application in load oscillations damping in platform of oil production, *Petróleo, Centro de Pesquisas (Cenpes)*, pp. 151–165. in Portuguese.

[10] Sausen, A. [2009]. *Mathematical modeling of a pipeline-separator system unde slug flow and control level considering an error-squared algorithm*, Phd thesis, Norwegian University of Science and Technology, Brazil. in Portuguese.

[11] Sausen, A. & Barros, P. R. [2007a]. Lyapunov stability analysis of the error-squared controller, *Dincon 2007 - 6th Brazilian Conference on Dynamics, Control and Their Applications*, São José do Rio Preto, Brasil, pp. 1–8.

[12] Sausen, A. & Barros, P. R. [2007b]. Properties and lyapunov stabilty of the error-squared controller, *SSSC07 - 3rd IFAC Symposium on System, Structure and Control*, Foz do Iguaçu, Brasil, pp. 1–6.

[13] Sausen, A. & Barros, P. R. [2008]. Modelo dinâmico simplificado para um sistema encanamento-*Riser*-separador considerando um regime de fluxo com golfadas, *Tendências em Matemática Aplicada e Computacional* pp. 341–350. in Portuguese.

[14] Shinskey, F. [1988]. *Process Control Systems: Application, Design, and Adjustment*, McGraw-Hill Book Company, New York.

[15] Storkaas, E. [2005]. *Stabilizing control and controllability: control solutions to avoid slug flow in pipeline-riser systems*, Phd thesis, Norwegian University of Science and Technology, Norwegian.

[16] Storkaas, E., Alstad, V. & Skogestad, S. [2001]. Stabilization of desired flow regimes in pipeline, *AIChE Annual Meeting*, number Paper 287d, Reno, Nevada.

[17] Storkaas, E. & Skogestad, S. [2002]. Stabilization of severe slugging based on a low-dimensional nonlinear model, *AIChE Annual Meeting*, number Paper 259e, Indianapolis, USA.

[18] Storkaas, E. & Skogestad, S. [2005]. Controllability analysis of an unstable, non-minimum phase process, *16th IFAC World Congress*, Prague, pp. 1–6.

[19] Storkaas, E. & Skogestad, S. [2006]. Controllability analysis of two-phase pipeline-riser systems at riser slugging conditions, *Control Enginnering Practice* pp. 567–581.

[20] Storkaas, E., Skogestad, S. & Godhan, J. M. [2003]. A low-dimensional dynamic model of severe slugging for control design and analysis, *11th International Conference on Multiphase flow (Multiphase03)*, San Remo, Italy, pp. 117–133.

[21] Tengesdal, J. O. [2002]. *Investigation of self-lifting concept for severe slugging elimination in deep-water pipeline/riser systems*, Phd thesis, The Pennsylvania State University, Pennsylvania.

[22] Thomas, P. [1999]. *Simulation of Industrial Processes for Control Engineers*, Butterworth heinemann.

Chapter 6

Integration of Seismic Information in Reservoir Models: Global Stochastic Inversion

Hugo Caetano

Additional information is available at the end of the chapter

http://dx.doi.org/10.5772/52308

1. Introduction

The stochastic models for reservoir properties characterization are a known important tool for reserve management, as well as reservoir quality and thickness that play a key role in deciding optimal well locations in any producing fields. Today's production reservoirs are each day more complex, and the majority of them, are of difficult access (off-shore), which in technical and cost terms, represents a lack of information.

In petroleum applications, stochastic modeling of internal properties (porosity and permeability), lithofacies and sand bodies of reservoirs, normally use core and log data which in the area provides detailed reservoir parameters, spatially it is limited to a few subsurface locations, scarce and expensive but it is reliable information.

The models created, with the lack of information, are models with great level of uncertainty. It is in this category of models that it is possible to find the stochastic simulation – Sequential Indicator Simulation to the morphological characterization of lithoclasses in [1], the Sequential Gaussian Simulation in [2] and recently, the Direct Sequential Simulation in [3].

The integration of different types of information in a unique and coherent stochastic model has been one of the most important, and still current, challenges of the geostatistical practice of modeling physical phenomena of natural resources and in order to make decisions regarding the development of well locations the geoscientists need to use all available data.

The recent trend of the scientific community regarding development and research for reservoir characterization is creating models which integrate other kind of information (secondary or auxiliary) normally available – the seismic information.

The seismic data which can cover the entire reservoir space has a high uncertainty given the quality and the vertical coarse resolution of seismic. This varies from 25 by 25 meters in horizontal and 1 to 4 milliseconds, in 3-D seismic acquisitions. This data sample is much coarser that the data measured in wells, which vary from some centimeters to a few feet. It is important information never the less, but in almost all applications the seismic data cannot have a direct link to the wells properties (lithology, porosity and permeability), and are difficult to use directly in the models one wishes to create.

The reservoir models based only in seismic information (3-D or 4-D), are normally limited to the structural information. This relationship derives from the major horizons and faults systems, interpreted in the coarse seismic, and it does not take in account the available well information, related to the internal characteristics of the reservoir (porosity, permeability and water saturation). On the other side, the characterization of reservoir models based only in the information of wells, like the recent geostatistical stochastic models, can have a great improvement by the integration of seismic information, which normally is available in the initial phases of prospecting and production.

The integration of these two types of information, with different special coverage and with totally different uncertainty levels is a challenge that even today dazzles the scientific community linked to the earth science modeling.

2. Objective

The main objective of this work is the development and implementation of a stochastic model algorithm for seismic inversion to improve reservoir characterization.

The methods of integration of seismic data can be roughly divided in two approaches. The methods that rely on a statistical relationship between seismic data and internal properties, or lithofacies, to characterize local distributions of these properties in any location of the reservoir by using, for example, co-simulations as [4,5], and others different approaches are posed as an inverse problem framework where the solution, the known amplitudes of seismic, are physically related with the unknown acoustic impedances (or porosity) by mean of a convolution model. Among them there is the so called geostatistical inversion in [6].

The major disadvantages and drawbacks of the direct models, are that the correlation found in the wells locations between the seismic and the internal properties (porosity and permeability), are normally low and sometimes spurious, condemning all the models from there to a great uncertainty.

From the recent deterministic inversion models, the great drawback, is caused by the lack of production of any measure of uncertainty and from not being robust (a great dependency from the seismic quality) and having little liability in almost all of the more complex reservoirs.

In 1993 Bortolli, launched the embryo of what is considered a liable alternative to the existing inverse models, the stochastic inversion. In this method the sequential gaussian simulation is used to transform, using an interactive process, each of the N verticals columns of the seismic cube.

Since then, the geostatistical seismic inversion has been a commonly used technique to incorporate seismic information in stochastic fine grids models.

Essentially, geostatistical inversion methods as in [7-9], perform a sequential approach in two steps:

i. Acoustic impedance values are simulated in each trace (a column of a 3D grid) based on well data and spatial continuity pattern as revealed by the variograms;
ii. The acoustic impedance values are transformed, by a convolution with a known estimated wavelet, into amplitudes giving rise to a synthetic seismogram that can be compared with the real seismic.

A "best" simulated trace is retained, based on the match of an objective function (function of the similitude between real seismic trace and seismogram), and another trace is visited to be simulated and transformed. The sequential process continues until all traces of acoustic impedances are simulated. In each step as long as the "best" transformed trace is accepted, the traces of simulated acoustic impedances are incorporated as "real" data for the next sequential simulation step. This can lead to artificially good matches in local areas where the bad quality of seismic prevails.

The base idea of this research work is precisely to incorporate stochastic simulation and co-simulation methodologies to conceive and implement a model of global seismic inversion and creating uncertainty linked to areas with different seismic quality.

The use of geostatistics for the creation and transformation of images (acoustic impedances) and the genetic algorithms for the modification and generation of better images, allow the convergence of the inverse process.

The methodology is proposed based on a global perturbation, instead of trace-by-trace, to reach the objective function of the match between synthetic seismogram and real seismic. Using the sequential simulation and co-simulation approaches it creates several realizations of the entire 3D cube of acoustic impedances that are simulated in a first step, instead of individual traces or cells.

After the convolution, local areas of best fit of the different images are selected and "merged" into a secondary image of a direct co-simulation in the next iteration.

The iterative and convergent process continues until a given match with an objective function is reached. Spatial dispersion and patterns of acoustic impedances imposed are reproduced at the final acoustic impedance cube.

As the iterative process is based on global simulations and co-simulations of impedances, there is no local imposing of artificial good fit, i.e. areas of bad seismic tend to remain with bad match coefficients, as it does not happens in most trace-by-trace approaches.

At each iterative step one knows how close is one given generated image from the objective, by the global and local correlation coefficients between the transformed traces and the real seismic traces. These correlation coefficients of different simulated images are used as the affinity criterion to create the next generation of images until it converges to a given predefined threshold.

In a last step, porosity images can be derived from the seismic impedances obtained by seismic inversion and the uncertainty derived from the seismic quality is assessed based on the quality of match between synthetic seismogram and real seismic.

For the case of characterization of the reservoir in terms of facies distribution several methods for the integration of the seismic data in facies models have been proposed, several of which rely on the construction of a facies probability cube by calibration of the seismic data with wells. If only post-stack seismic data is considered, is typically inverts the seismic amplitudes into a 3D acoustic impedance cube, and then converts this impedance into a 3D facies probability using a calibration method of choice.

This facies probability can serve as input of several well-known geostatistical algorithms to create a facies realization, such as the use the cube as locally varying mean on indicator kriging or the use of the tau model in [10], in a multi-point simulation to integrate the facies probability cube with spatial continuity information provided by a training image.

While this provides satisfactory results in most cases, the resulting facies realization does not necessarily match the original seismic amplitude from which the acoustic impedance was inverted. Indeed, if one would forward simulate, for example a 1D convolution on a single facies realizations, then this procedure does not guarantee that the forward simulated seismic matches the field amplitudes.

Another objective of this work is to present a geostatistical methodology, based on multipoint technique that generates facies realization compatible with the field seismic amplitude data, and therefore this new procedure has two main advantages, matching field seismic amplitude data in a physical sense, not merely in a probabilistic sense and using multi-point statistics, not just the two-point statistics (variogram).

3. Seismic inversion

The seismic data is the best source of spatially extensive measurement over the reservoir, but as explained previously, is very coarse vertically. From the wells we can use the acoustic impedance (AI) which is the result of multiplying the lateral interval velocity by the layer density, that later can be transformed in to seismic amplitude, this is possible because the difference between the different geological materials is linked to the response of the seismic signal.

The advantages of working with AI instead of the recorded seismic data are that is layer property rather than an interface property and hence more like geology and therefore it has a more physical meaning and during the inversion process all the well data is tied to the seismic data giving better understanding of the quality of both datasets.

So, the integration of the amplitude seismic data with the acoustic impedance data from wells requires some kind of transformation. There are two methods that transform one in to the other, an inverse method and a direct or forward method.

Both of them use wavelets for the transformation. These can be summarized as one-dimensional pulse and the link between seismic data and geology and they are a kind of impulse of energy that is created when a marine air gun or land dynamite source is released during the acquisition of seismic surveys.

The first method, "inverse process", transforms the amplitudes from seismic to acoustic impedance by removing the wavelet from the amplitudes and obtaining a model of the acoustic impedance. The problem of simultaneously inverting reservoir engineering and seismic data to estimate porosity and permeability involves complex processes such as fluid flow through porous media, and acoustic wave propagation and cannot be solved by linear inversion methods as [11]. On the other hand, the process of sending a wavelet into the earth and measure the reflection is called forward process. Here we need to know what it is in the subsurface in terms of geological models divided in layers and each characterized by his acoustic impedance. Forward modeling combines the sequence of differences in the acoustic impedances with a seismic pulse to obtain a synthetic amplitude trace.

In the forward model if we compare those sequences of synthetic amplitude variations with the real ones, we can identify and quantify the differences and try to minimize them.

3.1. Synthetic seismogram

It is known that the propagation of waves or sounds that pass through earth can be explained by an elastic wave equation. One simple, but powerful and commonly used approach for computing the seismic response of a certain earth model is the so-called convolutional model as in [12-14], which can be derived from an acoustic approximation of an elastic equation.

This convolutional model (equation 1) creates a synthetic seismic trace Sy(t), that is the result of convolving a seismic wavelet w(t) with a time series of reflectivity coefficients r(t), with the addition of a noise component n(t), as follows:

$$Sy(t) = w(t) * r(t) + n(t) \tag{1}$$

An even simpler assumption is to consider the noise component to be zero (equation 2), in which case the synthetic seismic trace is simply the convolution of a seismic wavelet with the earth's reflectivity:

$$Sy(t) = w(t) * r(t) \tag{2}$$

The reflection coefficient is one of the fundamental physical concepts in the seismic method. Basically, each reflection coefficient may be thought of as the response of the seismic wavelet to an acoustic impedance change within the earth and represents the percentage of the energy that is emitted compared to the one that is reflected

Mathematically, converting from acoustic impedance to reflectivity involves dividing the differences in the acoustic impedances by the sum of them. This gives the reflection coefficient at the boundary between the two layers.

Each layer can have different rock acoustic impedance, which is a function of two porosity-dependent rock properties: bulk density and velocity (equation 3). For instance, the P-wave acoustic impedance is given by:

$$AI(t) = density(t) * P\text{-}Velocity(t) \tag{3}$$

So the reflectivity coefficient (equation 4) can be computed from:

$$r_t = \frac{AI_{t+1} - AI_t}{AI_{t+1} + AI_t} \tag{4}$$

where AI represents the acoustic impedance and the subscripts t and t+1 refer to two subsequent layers in the stratigraphic column.

The wavelet is usually extracted from the seismic survey through deconvolution (the methodology of extraction the wavelet from the seismic signal). During the deconvolution, the wavelet extraction does not take in account the low frequency of the earth model, this represents the compaction of the porous media in depth, which in some seismic inversion algorithm and in the inverse modeling, are ignored or only take in account adjacent geology units, not the full vertical analysis.

In this work, the problem is resolved by the algorithm of generation of the acoustic impedance model, consider the trend that are represented in the wells log data, by creating initially a model only conditioned by the wells data (AI).

3.2. Correlation coefficient comparison

Using the convolution algorithm described previously, one can forward an entire model (or cube if one considers a model as a cartesian grid) of AI created only by using the wells and its spatial continuity to a synthetic seismic amplitude cube that can be compared with the real amplitude seismic cube. This comparison will give the mismatch between those two cubes of amplitudes.

The mismatch between the synthetic and the field amplitudes is calculated as a simple co-located correlation coefficient as in [15-17]. Alternatively, this mismatch could be calculated in the form of a least-square difference between both amplitude cubes.

In a first step, each of the N synthetic seismic amplitude cubes are divided into layers (figure 1).The choice of number of layers is a tuning parameter of the algorithm.

The amount of mismatch between each pair of columns of the synthetic and the real amplitude is calculated and stored in a new cube (figure 1), named the mismatch cube. Essentially, this new cube provides information on which locations in the reservoir are fitting the seismic and which need improvement.

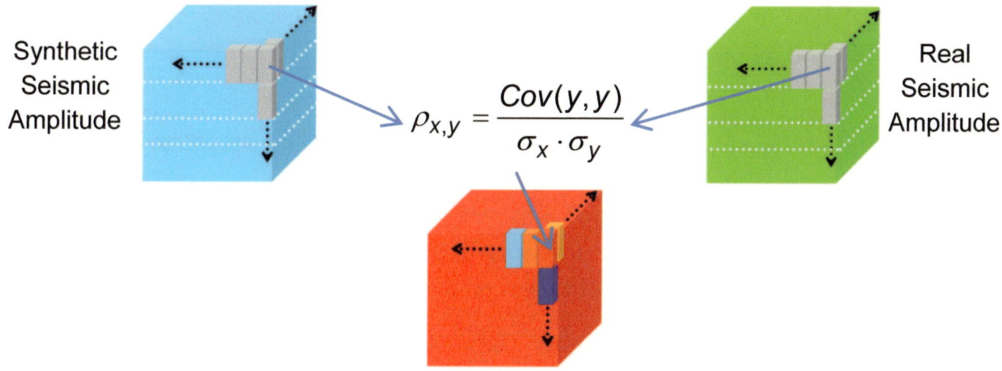

Figure 1. Calculus of the mismatch cube.

A mismatch cube is calculated for each of the simulated models using their corresponding synthetic seismic models.

3.3. Co-generation of acoustic images

The generation of an acoustic impedance model is the main objective of this work, but until an optimized one is created, it is submitted to a convergence genetic modification.

These images are generated through stochastic simulation, i.e. generation of different models with the same probabilistic and spatial distribution. This means that if the parameters that are used to the creation of the models are too tight, the produced images are almost identical, if the parameters allow more freedom, these models are totally different.

Using this idea, the proposed methodology uses two approaches for image generation, two different programs are used. The first one is the DSS (Direct Sequential Simulation) for the generation of the first unconditioned images and then for the convergence is used the Co-DSS (Direct Sequential Co-Simulation) that uses a soft or secondary data to simulation conditioning as in [3,18,19].

The second algorithm is the Snesim (Single Normal Equation SIMulator), a multi point simulation algorithm described in [20,21]. This version is used to create the first unconditional facies models. A second version with modification in [2], uses the influence of secondary data to create the models that are conditioned to more information. Both these programs use secondary data for the co-simulations, and this secondary or soft data is the one that makes the algorithm converge.

For the use of Direct Sequential Co-Simulation for global transformation of images as in [17], let us consider that one wishes to obtain a transformed image $Z_t(x)$, based on a set of N_i images $Z_1(x)$, $Z_2(x)$,...$Z_{Ni}(x)$, with the same spatial dispersion statistics (e.g. correlogram or variogram and global histogram): $C_1(h)$, $\gamma_1(h)$, $F_{z1}(z)$.

Direct co-simulation of $Z_t(x)$, having $Z_1(x)$, $Z_2(x)$,...$Z_{Ni}(x)$ as auxiliary variables, can be applied as in [3]. The collocated cokriging estimator (equation 5) of $Z_t(x)$ becomes:

$$Z_t(x_0)^* - m_t(x_0) = \sum_{\alpha} \lambda_\alpha(x_0)\left[Z_t(x\alpha) - m_t(x_\alpha)\right] + \sum_{i=1}^{Ni} \lambda_i(x_0)\left[Z_i(x_0) - m_i(x_0)\right] \tag{5}$$

Since the models $\gamma_i(h)$, i=1, N_i, and $\gamma_t(h)$ are the same, the application of Markov approximation is, in this case, quite appropriated, i.e., the co-regionalization models are totally defined with the correlation coefficients $\rho_{t,i}(0)$ between $Z_t(x)$ and $Z_i(x)$.

The affinity of the transformed image $Z_t(x)$ with the multiple images $Z_i(x)$ are determined by the correlation coefficients $\rho_{t,i}(0)$. Hence, one can select the images with which characteristics we wish to "preserve" in the transformed image $Z_t(x)$. So, a local screening effect approximation can be done, by assuming that to estimate $Z_t(x_0)$ (equation 5) the collocated value $Z_i(x_0)$ of a specific image $Z_i(x)$, with the highest correlation coefficient $\rho_{t,i}(0)$, screens out the influence of the effect of remaining collocated values $Z_j(x_0)$, $j \neq i$. Hence, equation 6 can be written with just one auxiliary variable: the "best" at location x_0:

$$Z_t(x_0)^* - m_t(x_0) = \sum_{\alpha} \lambda_\alpha(x_0)\left[Z_t(x\alpha) - m_t(x_\alpha)\right] + \lambda_i(x_0)\left[Z_i(x_0) - m_i(x_0)\right] \tag{6}$$

With the local screening effect, N_i images $Z_i(x)$ give rise to just one auxiliary variable. The N_i images are merged in on a single image based on the local correlation coefficient criterion.

Basically, the correlation or least mismatch of each of the previously simulated images at each x_0 is converted to a single image where the best of each x_0 is selected, this way.

In practice, the algorithm chooses in each x_0 of the simulated acoustic impedance model the best genes and the correlation that those genes have. Based on figure 2, the process is explained as follows:

The process starts by analyzing each of the mismatch cubes that were created previously, and in each position of the cubes (x_0) it compares which has the higher value of correlation or lower mismatch. In this case the example used is the correlation, so, the points 1a and 1b are compared, and since it is the 1a that has the higher value of correlation, that value is copied (1c) to the Best Correlation Cube.

At the same time, since it was from the first simulation that has better coefficient of correlation (CC) with the real seismic, another cube is created with the correspondent acoustic impedance (AI) value (1d). Next, it starts comparing the next set of values from the CC cubes, and the process is the same. With a comparison between the sets 2a and 2b, and since in this case it is the second cube that has the higher value of cc, it is this values that is copied to the Best CC cube (2c). The same is done from the AI cube of this simulation, copying the acoustic impedance values to the Best AI (2d). This process continues until all the N simulated models have been analyzed and to all the values of the cubes. Through this way two new cubes are created, one with the best genes from each acoustic impedance model (Best AI) generated previously and another cube with the confidence factor of each part (Best CC).

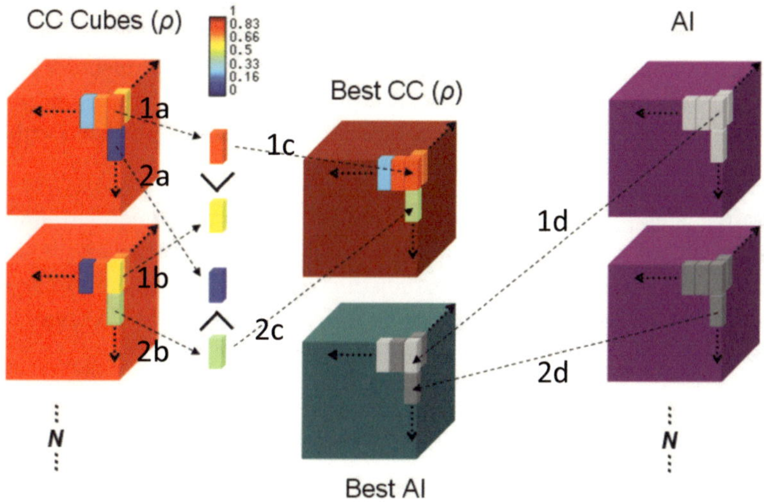

Figure 2. Process of creating "Best Correlation Cube" and "Best Acoustic Impedance cube"

Since the process is an iterative one, in the very first steps of the iterative process, the secondary image (Best AI) do not have the spatial continuity pattern of the primary (simulated AI), as it results from a composition of different parts of a set of simulated images. As the process continues, the secondary image tends to have the same spatial pattern of generated images by co-simulation, because the correlation values are becoming higher and the freedom of the co-simulation is diminishing. Finally these two new data sets are used as soft data for the next iteration.

4. Algorithm description using direct sequential simulation

The proposed stochastic inversion algorithm is settled with the following key ideas in mind: generation of stochastic images, transformation of the images in synthetic seismograms, chose and keep the best "genes" from each of the images and then use them through the exercise of the genetics algorithm formalism and the stochastic co-simulation to create a new generation of images and the convergence of the process. It can be summarized in the following steps;

i. Generate a set of initial 3D images of acoustic impedances by using direct sequential simulation. Instead of individual traces of cells;
ii. Create the synthetic seismogram of amplitudes, by convolving the reflectivity, derived from acoustic impedances, with a known wavelet;
iii. Evaluate the match of the synthetic seismograms, of entire 3D image, and the real seismic by computing, for example local correlation coefficients;
iv. Ranking the "best" images based on the match (e.g. the average value or a percentile of correlation coefficients for the entire image). From them, one selects the best parts (the columns or the horizons with the best correlation coefficient) of each image. Compose one auxiliary image with the selected "best" parts, for the next simulation step;
v. Generate a new set of images, by direct sequential co-simulation, using the best locals correlation as the local co-regionalization model and return to step ii) starting an iterative process that will end when the match between the synthetic seismogram and the real seismic is considered satisfactory or until a given threshold of the objective function is reached;
vi. The last step of the process is the transformation of the optimized cube of acoustic impedance in internal reservoir characteristics.

At each iterative step one knows how closer is one given generated image from the objective, by the global and local correlation coefficients between the synthetic seismogram and the real seismic. These correlation coefficients of different simulated images are used as the affinity criterion to create the next generation of images until it converges to a given predefined threshold. A simplified diagram is show in figure 3.

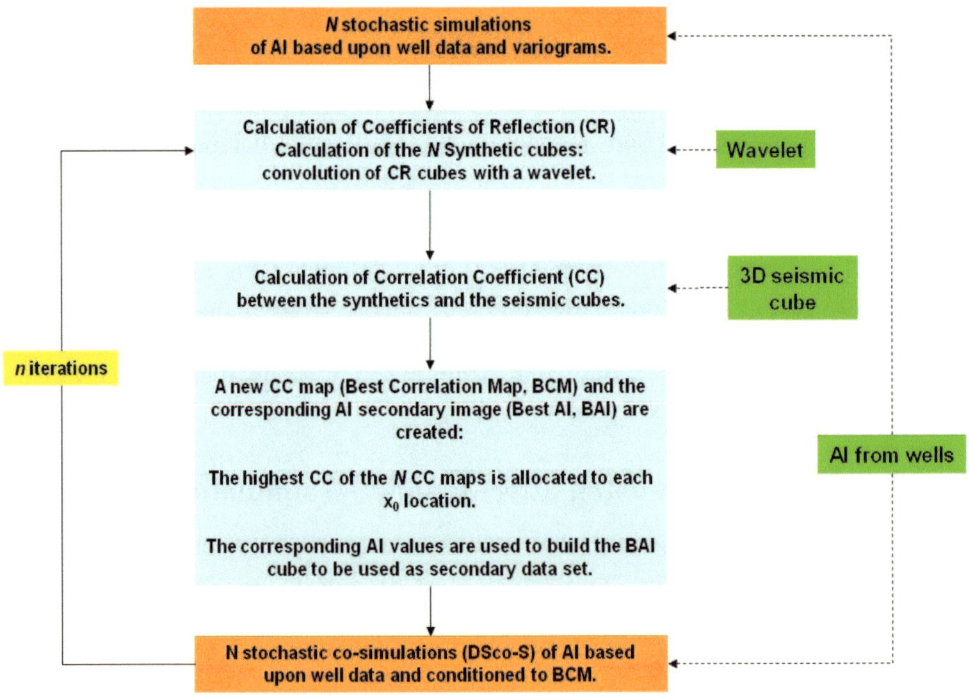

Figure 3. Diagram of the proposed algorithm

4.1. Case study

A case study of a Middle East reservoir will be presented but only a small part of the full reservoir is studied due to data confidentiality and the coordinates presented are modified.

The field described in [22,23] is a carbonate reservoir with a deposition geology that holds in some zones a strong internal geometry with clinoforms. These are sedimentary deposits that have a sigmoidal or S shape. They can range in size from meter, like sand dunes, to kilometers and can grow horizontally in response to sediment supply and physical limits on sediment accumulation.

This type of geological phenomenon increases the complexity of the internal reservoir, making the geophysics interpretation of seismic a very difficult job, mainly when the resolution of acquisition is very coarse. It also causes a great impact in oil production in wells and reservoir characterization and modeling.

From the calibration of the acoustic impedance data of the wells with the seismic, a wavelet was extracted and there were 19 wells in the study area but only two were used, because only these two have the velocity log, and without the velocity log the acoustic impedance data cannot be calculated. On those wells the acoustic impedance was determined, then calibrated with the seismic and finally upscaled to fit the seismic scale of 4 ms.

The convolution for well 11 is show in figure 4.

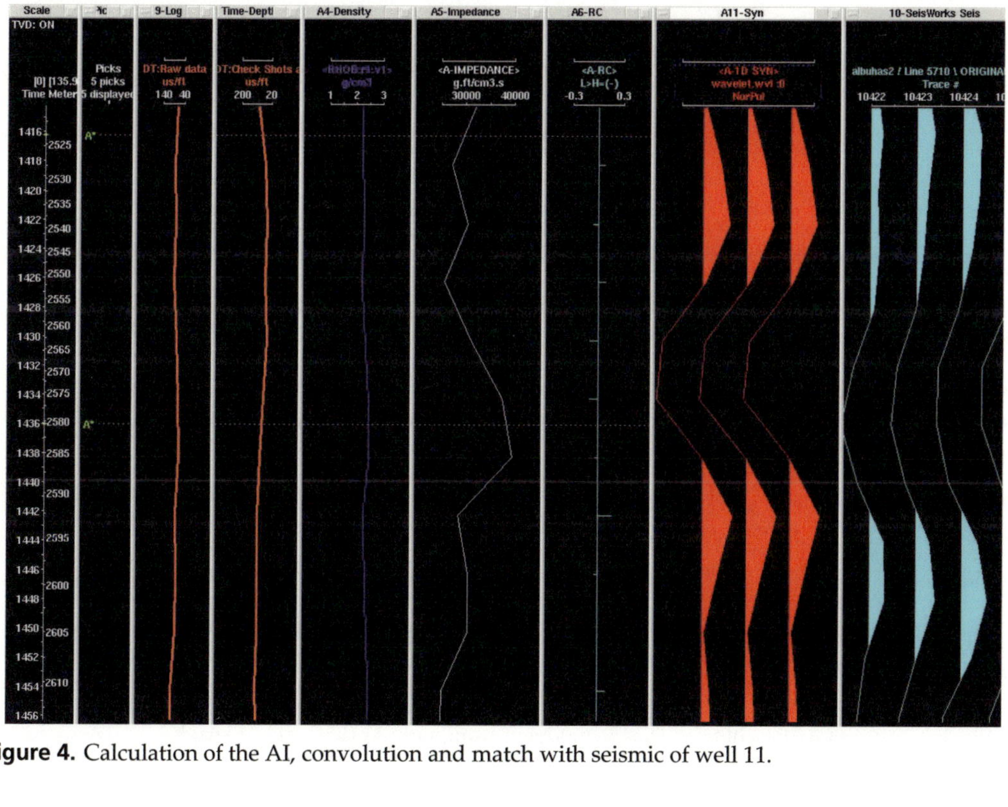

Figure 4. Calculation of the AI, convolution and match with seismic of well 11.

As one can notice, there is a very good match with the synthetic amplitude (red line on right side) calculated with the acoustic impedance (white line in the middle) from the wells and the real seismic amplitude extracted from the seismic cube (cyan line in the right side).

4.2. Results

The first sets of 32 images of acoustic impedances are generated with the direct sequential simulation conditioned to well data (AI) and the chosen variogram model. In the first iteration, the generated acoustic models (figure 5) are not constrained to any soft data, so

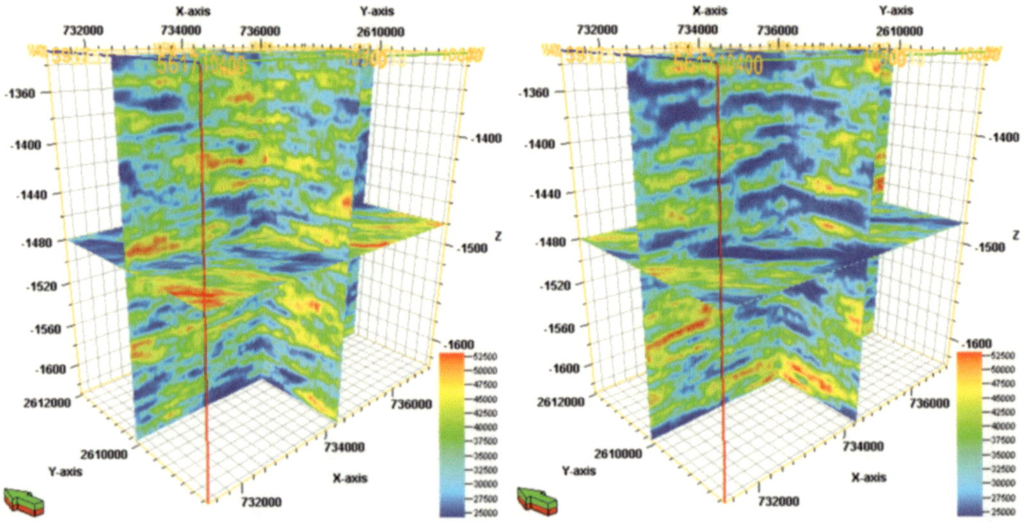

Figure 5. Acoustic Impedance model of simulation #1 (left) and #15 (right).

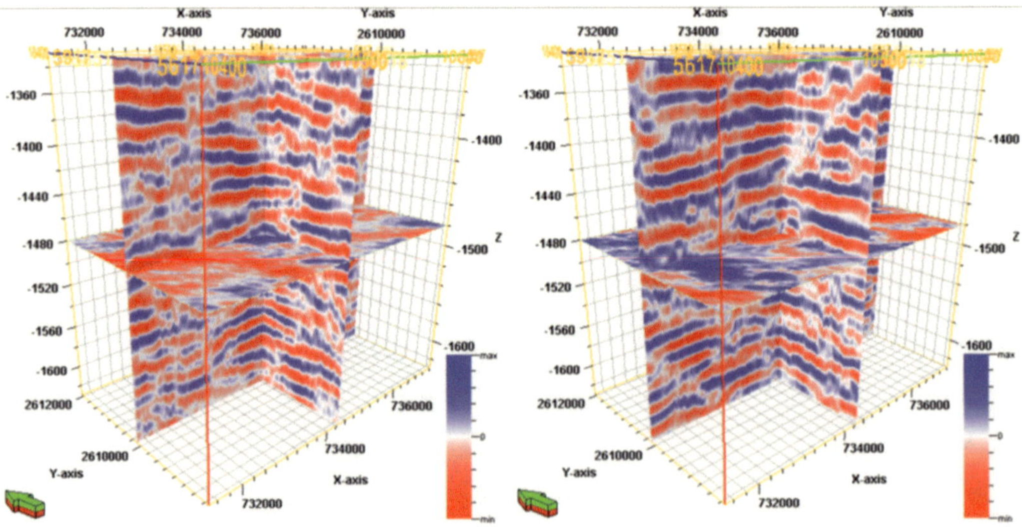

Figure 6. Correspondent Synthetic Seismic model of simulation #1 (left) and #15 (right).

only the wells are reproduced. This causes a high variability between the synthetic models (figure 6) and wide range standard deviation when calculated using all 32 simulations.

As it can be noticed, the two models have a totally different spatial distribution, although the histograms and variograms are the same.

That different spatial distribution is visible in the correlation cubes between the synthetic models and the real seismic (figure 7).

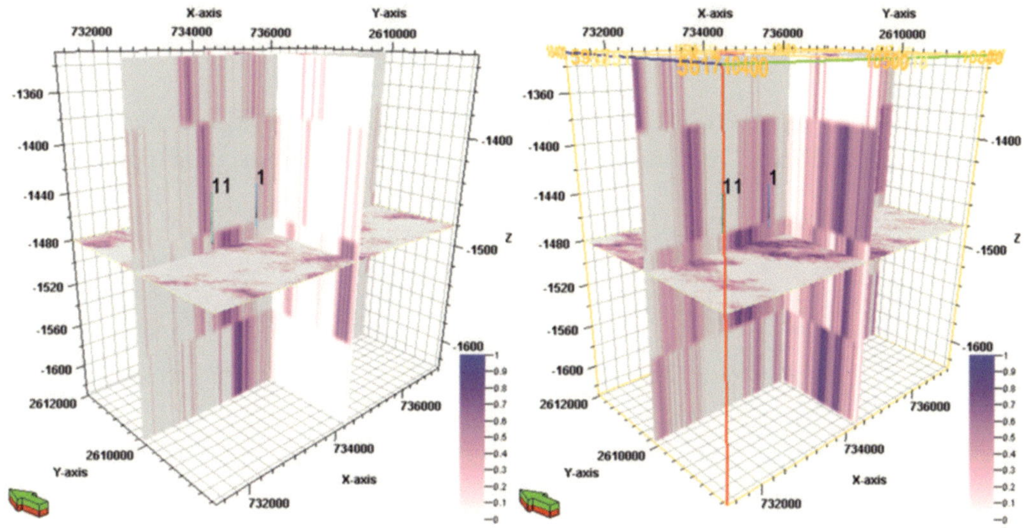

Figure 7. Correlation cubes between the synthetic seismic model and the real seismic, of simulation #1 (left) and #15 (right).

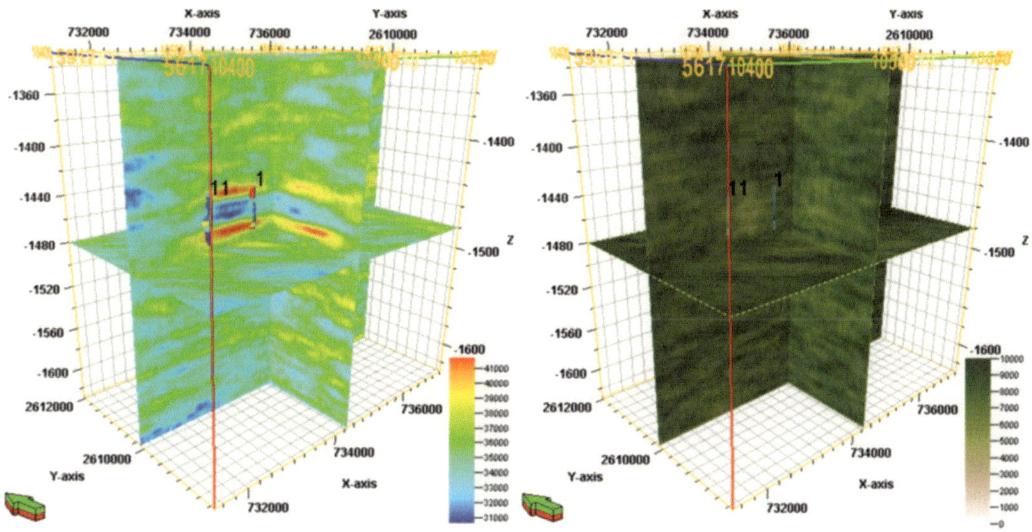

Figure 8. Average (left) and Standard Deviation (right) of acoustic impedance of all 32 simulations of this iteration.

To confirm this variability, the average cube and standard deviation (figure 8) of all 32 simulations were determined and the influence of the wells in the average of the unconditioned simulations is highly visible principally around well 1, as the variogram is reproducing the well log spatial variability.

As perceived in the average model, the small scale variability has disappeared, however this cube can be considered the Low Frequency Model, because it reproduces the main trends that are represented in the wells logs. Also noticeable the practically constant value of standard deviation, representing the variability of unconditional stochastic simulations, except the small decline of values in the middle of the figure, which is the influence of the well 1.

Still, these first 32 simulations were used to build the "Best Correlation Cube" and "Best Acoustic Impedance cube" that will be used as soft data for the next iteration (figure 9).

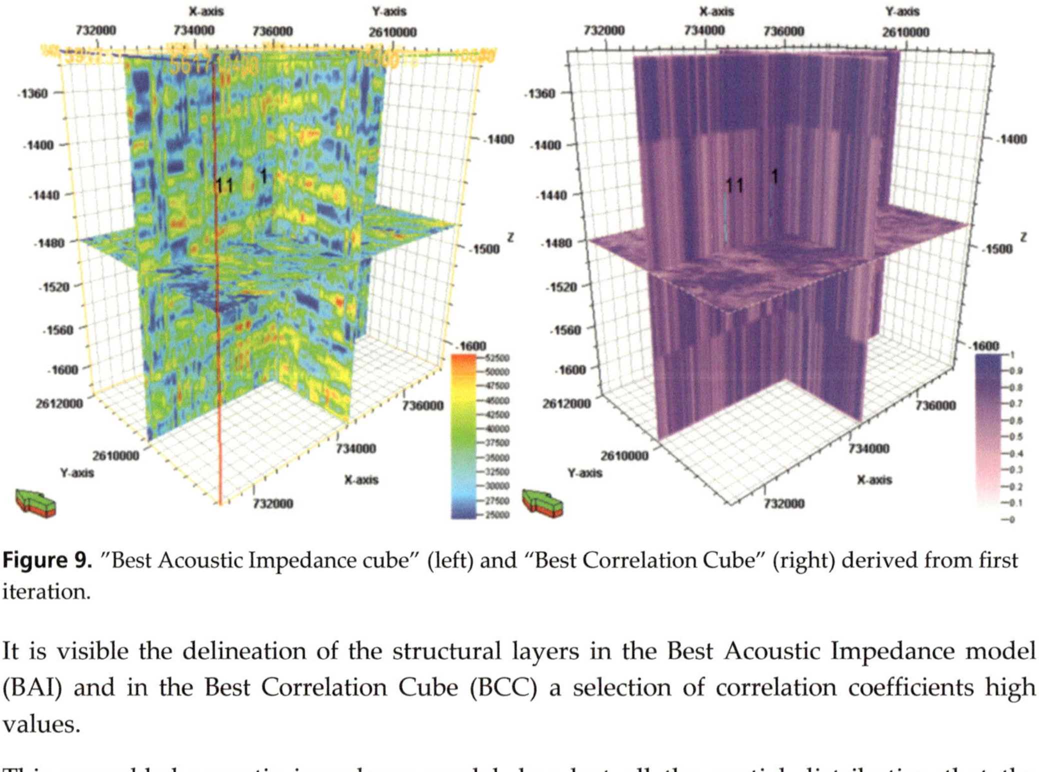

Figure 9. "Best Acoustic Impedance cube" (left) and "Best Correlation Cube" (right) derived from first iteration.

It is visible the delineation of the structural layers in the Best Acoustic Impedance model (BAI) and in the Best Correlation Cube (BCC) a selection of correlation coefficients high values.

This assembled acoustic impedance model, has lost all the spatial distribution that the original acoustic impedance models had (figure 5), but this is a simple intermediate result and not the final one since these will be used as secondary or soft data for the next iteration that will impose the variogram spatial distribution and wells global histogram.

The algorithm has made six iterations, one of them, the first, was unconditional to any secondary or soft data, only the last five had the Best Correlation Cube and Best Acoustic Impedance cube as secondary or soft data imposition.

Since the algorithm will always choose the best genes from each iteration, the patterns that are in the real seismic will start to become more visible in each iteration and the correlation will became higher and more continuous in all cube positions.

The convergence of the process is inevitable until a local maximum correlation is attained (figure 10). This maximum does not represent the best that the original seismic can produce, but rather a simulation that the entire process has created. If one has run the same process with the same data but with a different seed for the generation of the acoustic impedance model the result could be a different one, but not so totally different.

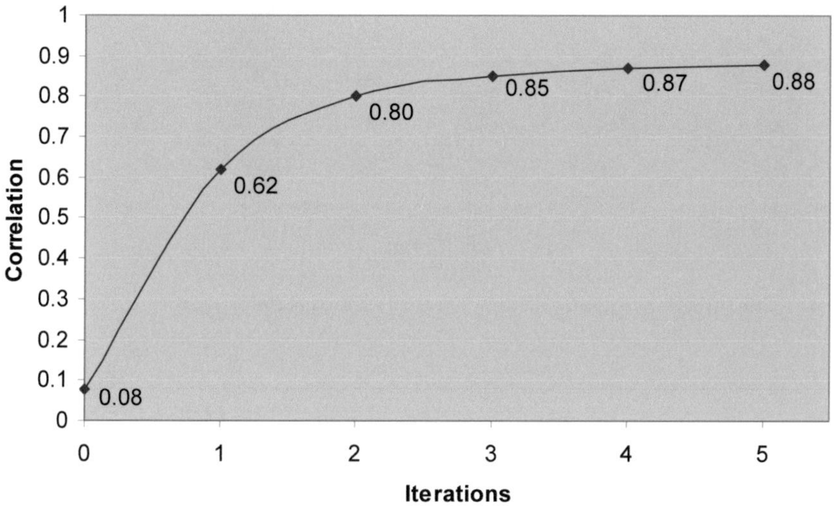

Figure 10. Convergence progress of the algorithm.

The 32 images of acoustic impedances of iteration 5 (considering that the first iteration is called 0, because it is an unconditional to soft data) are generated with the direct sequential co-simulation conditioned to well data (acoustic impedances), the chosen variogram model and the soft or secondary data (BAI and BCC) of the forth iteration.

One can clearly see the fast convergence from iteration zero to iteration 1 and afterwards the process starts to stabilize. The algorithm chooses the parts that have higher correlation values in the end of iteration 0 and after iteration 1, almost every simulation has its correlation values around 1, making the selection of each part of simulation, a very detailed event (sometimes in the third of forth decimal of the correlation value).

The obtained final results demonstrate that the conditional data is imposing a very strong effect in all 32 simulations. The variability of different models generated with different seeds is now almost none existing (figure 11 and 12) and they practically look the same.

As it can be noticed the two models have an almost equally spatial distribution, but some differences can be found, since they are two independent realizations.

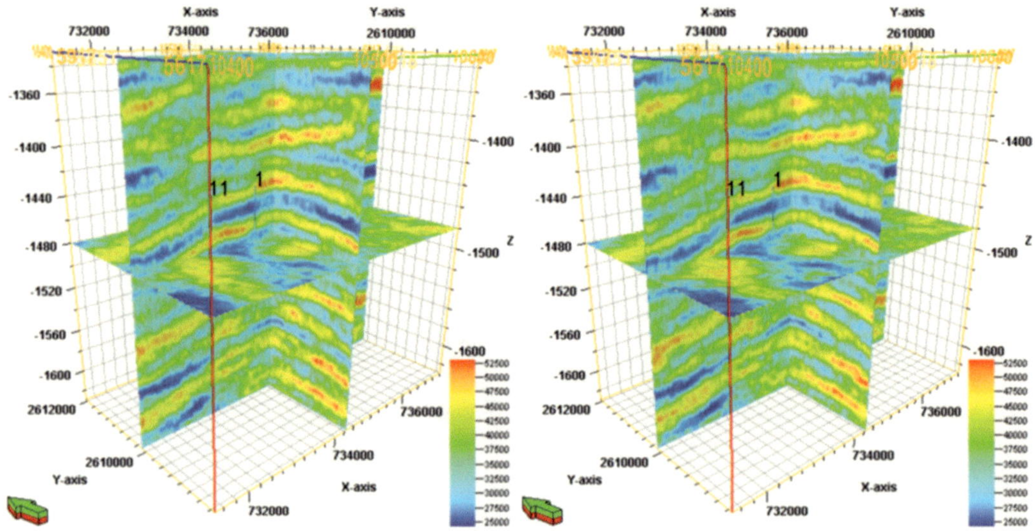

Figure 11. Acoustic Impedance model of simulation #3 (left) and #28 (right).

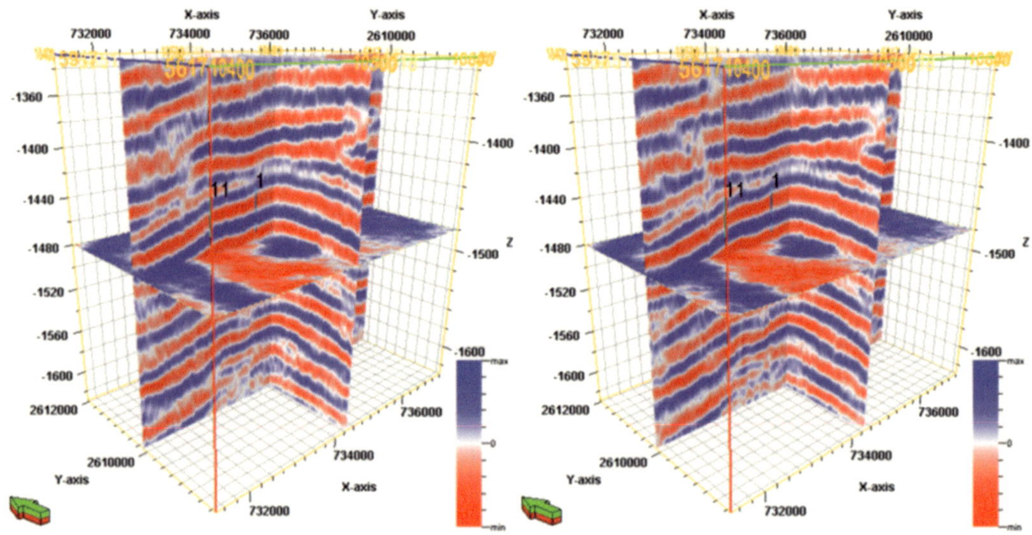

Figure 12. Correspondent Synthetic Seismic model of simulation #3 (left) and #28 (right).

Those small differences are easier to distinguish in the correlation cubes between the synthetic models and the real seismic (figure 13), since the correlation coefficient is very sensible to little variations in patterns.

In these examples the layer set sizes were big enough to show the differences.

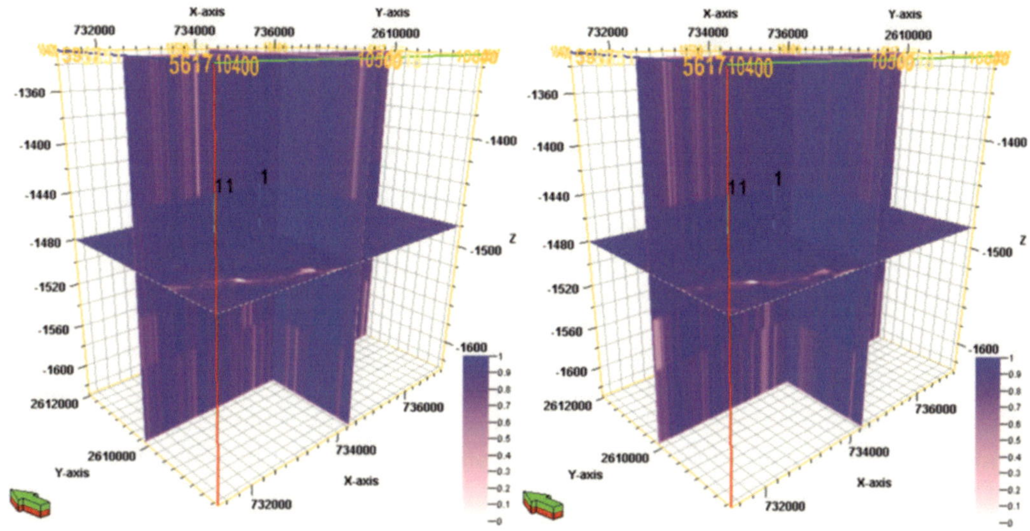

Figure 13. Correlation cubes between the synthetic seismic model and the real seismic, of simulation 3 (left) and 28 (right).

To confirm this lack of variability (figure 14), the average cube is almost equal to a simulation, and validated by the lower values of standard deviation of all 32 simulations.

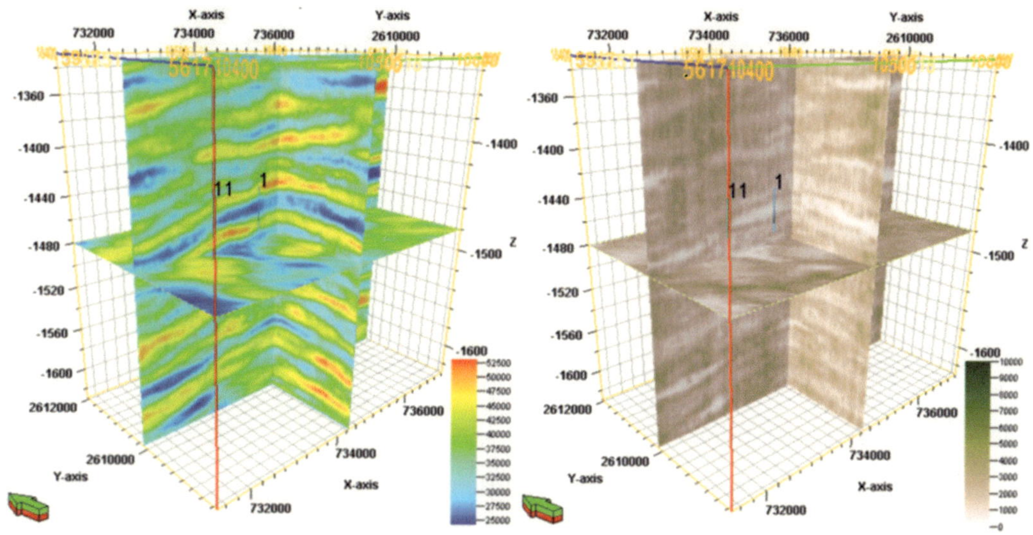

Figure 14. Average and standard deviation of acoustic impedance of all 32 simulations of this iteration.

In figure 14 the same color scale for standard deviation, was used, for comparison with the standard deviation of the first iteration model (figure 8).

The values of standard deviation for the final iteration only vary from 0 to 6900, with an average of 1600 which is a reduction of almost 80% in variability, comparing with the variation between 0 and 11500 and average of 8000 of the first iteration.

The influence of the wells is no longer visible because all data of wells are integrated in the full model.

4.3. Remarks

In spite of the scarcity of the log data, the proposed method achieved extremely good results. In case of data abundance, if the data is not well calibrated it could compromise the quality of the convergence and the influence of the not calibrated wells in the final model would be noticeable.

In this case, both synthetic seismic data and acoustic impedance cube captured the main geologic features of these complex reservoirs, noticeable in the correlation coefficients between the seismic and the synthetic amplitudes. The quality of the seismic data takes a minor role since the method overcomes the situation of imposition of artificial correlations as it happens in the standard methods;

Since the co-simulation of the impedances uses a local coefficient correlation, it is possible to compute the local uncertainty associated to the seismic acoustic impedances;

The uncertainty of the seismic acoustic impedance could be used to access the uncertainty associated with the porosity model, as presented in [24].

Tests prove that the variation of the final correlation coefficient is about 2% with the modification of the initial seed, it means that others parameters such as the number of layers and the size of it, as other parameters can be optimized to produce better results. But these results are more difficult to reach when the complexity of the geology and the structural model became more elaborated.

To handle different geology scenarios such channels or specific shape reservoirs, an adapted approach is proposed in the next part of this work.

5. Multipoint statistics

The objective of this work is to build a reservoir model with multiple alternatives, thereby assessing uncertainty about the reservoir, integrating information from different sources obtained at different resolutions:

- Well-data which is sparse but of high resolution, in the scale of a foot;
- Seismic data which is exhaustive but of much lower resolution, in the scale of 10's feet in the vertical direction;
- Conceptual geological models, which could quantify reservoir heterogeneity from the layer scale to the basin scale.

This last point has been possible using a variogram in algorithms such as DSS and Co-DSS, previously described, which allow integration of well and seismic data using a pixel-based approach. Those variogram based models are inadequate in integrating geological concepts since the variogram is too limited in capturing complex geological heterogeneity and is a two-point statistics that poorly reflects the geologist's prior conceptual vision of the reservoir architecture, e.g., sand channels as in [20].

Integration of geological information beyond two-point variogram reproduction becomes critical in order to quantify more accurately heterogeneity and assess realistically the uncertainty of model description.

A solution was initiated by practitioners from the oil industry in Norway, where Boolean objects-based algorithms were introduced in the late 1980's to simulate random geometry as in [25,26]. These parametric shapes, such as sinusoidal channel or ellipsoidal lenses, are placed in simulated volume. Through an iterative process this shape is changed, displaced or removed to fit the conditioning statistics and local data.

The simulated objects finally resemble the geologist drawings or photographs of present day depositions

Passed the enthusiasm, the limitation of objects-based simulation algorithms became obvious, this iterative, perturbation type, algorithm for data conditioning did not converge in the presence of dense hard data or could not account for diverse data types, such as seismic used as soft data.

Also the limitation of not simulating continuous variables, time and CPU demanding in large 3D cases, enroll on the drawbacks of the methodology.

Strebelle in [20], following the works of Srivastava in [27] and Caers in [28] has proposed an alternative approach of Multipoint statistics that combines the easy conditioning of pixel-based algorithms with the ability to reproduce "shapes" of object-based techniques, without relying on excessive CPU demand.

Multipoint statistics uses a training image instead of a variogram to account for geological information. The training image describes the geometrical facies patterns believed to represent the subsurface.

Training images does not need to carry any local information of the actual reservoir, they only reflect a prior geological/structural concept.

Basically, multipoint statistics consists of extracting patterns from the training image, and anchoring them to local data, i.e. well logs and seismic data. Several training images corresponding to alternative geological interpretations can be used to account for the uncertainty about the reservoir architecture.

Training images, identical to the variogram, can be a questionable subject when the model to be generated has few hard data or no pre-conceptual model has been studied by geologists or geomodelers.

Using this new tool a new objective is to illustrate the multipoint statistical methodology to seismic inversion.

5.1. Seismic inversion using multipoint statistics

For the case study, the Stanford VI synthetic reservoir dataset described in [29], was used. This synthetic reservoir (figure 15), consists of meandering channels, sand vs mud system, and each facies were populated with rock impedance, using sequential simulation.

To mimic a seismic inversion, this high resolution rock impedance model were smoothed by a low-pass filter to obtain a typical acoustic impedance response and then convoluted to an amplitude model that will be considered as reference seismic amplitude.

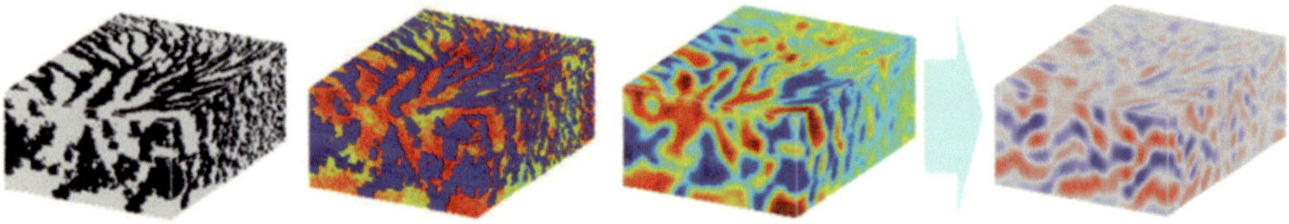

Figure 15. Facies model (left), high resolution rock impedance (middle left), smothered rock impedance (middle right) and forwarded seismic synthetic amplitude (right)

5.2. Probabilistic approach

For comparison of the methodology, is presented the application of a traditional probabilistic modeling on the Stanford VI acoustic impedance, i.e. calibrate acoustic impedance into a 3D facies probability cube, and then use it as soft data constraint for multiple-point geostatistical simulation.

Several techniques exist to calibrate one or more seismic attributes with well data, such as PCA, ANN, etc., one use a simple Bayes' approach (equation 7) as documented in [30]. In a Bayesian method one uses the histogram of impedance for each facies denoted as:

$$P\left(AI \mid facies_f\right), f = \{1,2,...\} \tag{7}$$

Using Bayes' rule one can calculate the probability for each facies for given impedance values as (equation 8):

$$P\left(facies_f \mid AI\right) = \frac{P(AI \mid facies_f) \bullet P(facies_f)}{P(AI)} \tag{8}$$

where $P(facies_f)$ is the global proportion of $facies_f$ and P(AI) is derived from the histogram of the impedance values. The final result is a cube of probabilities for each facies type based on the acoustic impedance cube. In figure 16, one can see the result of the application of this method to the case study data.

This probability data can be used by different geostatistical algorithms to create a facies model. In our case the simulated facies are obtained with the multi-point method snesim, conditioned to the probability cubes previously calculated, to the training image and to the rotation and affinity cubes.

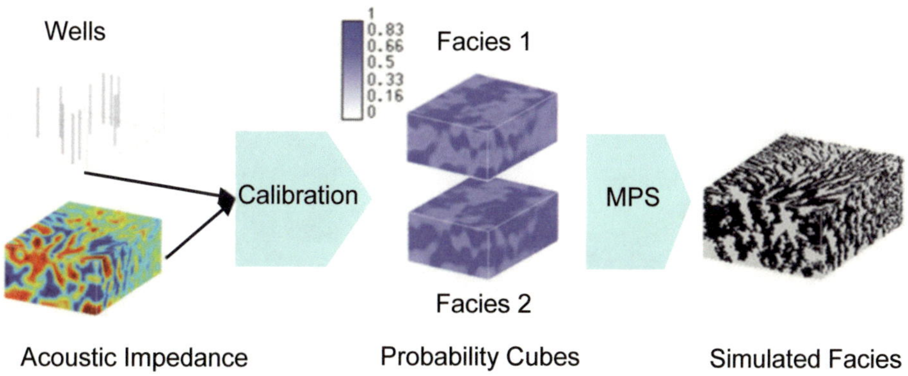

Figure 16. Result of the Bayesian approach for the calculus of the probabilities cubes for case study and the simulated facies model.

To verify our hypothesis that this facies model does not match the field amplitude, the facies model is forwarded simulated to a synthetic amplitude dataset. We assumed the ideal situation where an exact forward model was available. Figure 17 confirm that the forward modeled amplitude does not match the field data. In fact, the co-located correlation coefficient is 0.20.

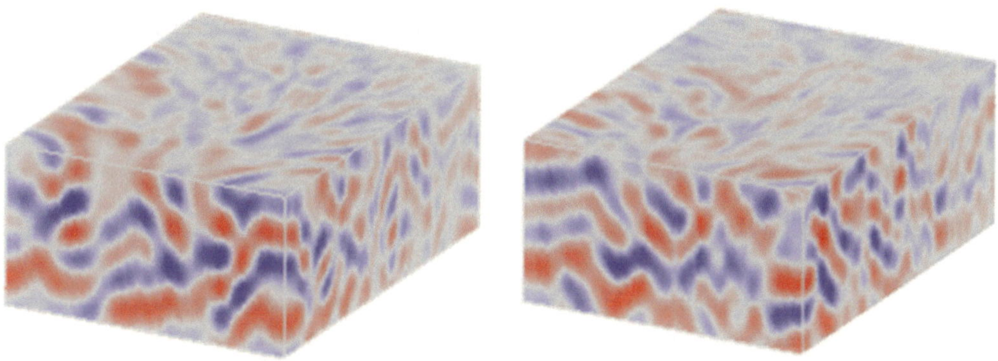

Figure 17. "Real" seismic amplitude (left) *vs* forward modeled amplitude derived from the simulated facies model using a traditional probabilistic approach (right).

5.3. Algorithm description using multipoint statistics

For the case of using multi-point statistics, the seismic inversion algorithm has gone through some changes, manly caused by the simulation algorithm that does not creates acoustic impedance models directly, but facies models, that can be populated with acoustic impedance values later.

So the method is initialized by simulating N facies models only conditioned to the wells and training images using snesim, in this case, additional channel azimuth and affinity (as interpreted from seismic or geological understanding) constraints are enforced (figure 18).

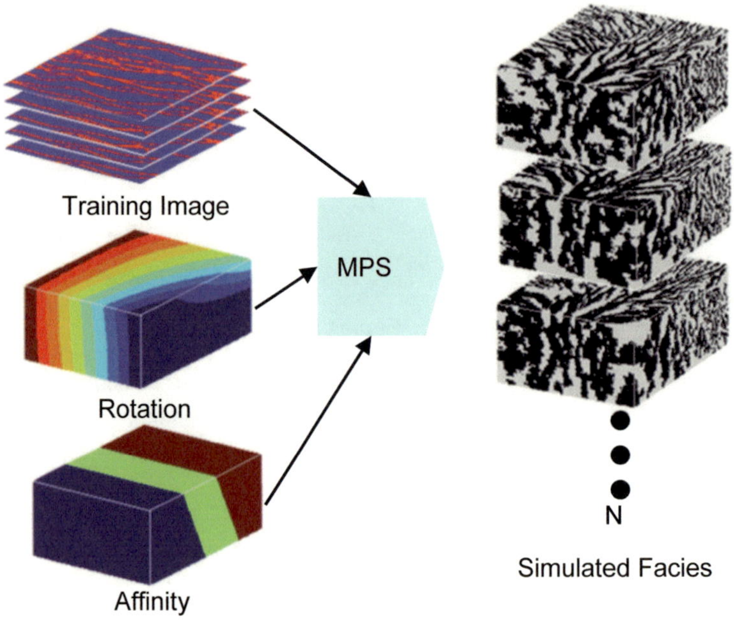

Figure 18. Initial simulations conditioned only to the training image and rotation/affinity.

The facies models are then populated with acoustic impedances, converted in synthetic seismograms and chose and keep the best "genes" from each of the images (see figure 19). In the next step there is also a modification, since one cannot use directly the correlation values in the snesim simulator. So a modification of the probability perturbation method as in [31] is used to create a probability for each facies.

This facies probabilities cube is then used as soft constraint to generate the next set of N cubes. This is done using Journel's tau-model as in [10], to integrate probabilistic information from various sources. The process converges as the previous algorithm and it can be summarized in the following steps:

i. Generate a set of initial 3D images of facies by using snesim.
ii. Populate the facies models with acoustic impedances.
iii. Create the synthetic seismogram of amplitudes, by convolving the reflectivity, derived from acoustic impedances, with a known wavelet.
iv. Evaluate the match of the synthetic seismograms, of entire 3D image, and the real seismic by computing, for example local correlation coefficients.
v. Ranking the "best" images based on the match (e.g. the average value or a percentile of correlation coefficients for the entire image). From it, one selects the best parts (the columns or the horizons with the best correlation coefficient) of each image. Compose one auxiliary image with the selected "best" parts, for the next simulation step.
vi. Transform the cube with the best parts and the corresponded best values to a probability cube to each facies to be simulated.

vii. Generate a new set of images using snesim adaptation to integrate secondary information, with the probability cubes used as the local co-regionalization model and return to step ii) starting an iterative process that will end when the match between the synthetic seismogram and the real seismic is considered satisfactory or until a given threshold of the objective function is reached.

viii. The last step of the process can be the transformation of the optimized facies cube in acoustic impedance, porosity or any internal reservoir characteristics.

As noted in previously algorithm, in step vi), in the very first steps of the iterative process, the probability data can not have the spatial continuity pattern of the primary simulated facies, as it results from a composition of different parts of a set of simulated images. As the process continues, that soft secondary image (probability cubes) tend to have the same spatial pattern of generated images by co-simulation, i.e. the imposed training image altered by the affinity and azimuth

5.4. Co-generation of acoustic images with multipoint statistics

The major change made for the adaptation to MPS, is that the acoustic impedance values are estimated or populated in the simulated facies model. This can cause some reservations but if one considered that the main objective is comparing the reflection coefficients between the real and the simulated seismic amplitude, those reflection coefficients are more accentuated in the change of terrain type, and those can be considered a channel or crevasse. Inside the channels the rock type does not vary too much and that is visible in the seismic profiles by the presence of low amplitude seismic. So this difference in the algorithm can be considered a valid one.

The second difference is that to co-simulate a facies model, one can not use acoustic impedance values (a continuous variable) as soft data, so instead of choosing from the simulated acoustics model, the algorithm build the Best Facies cube from the simulated facies models (figure 19).

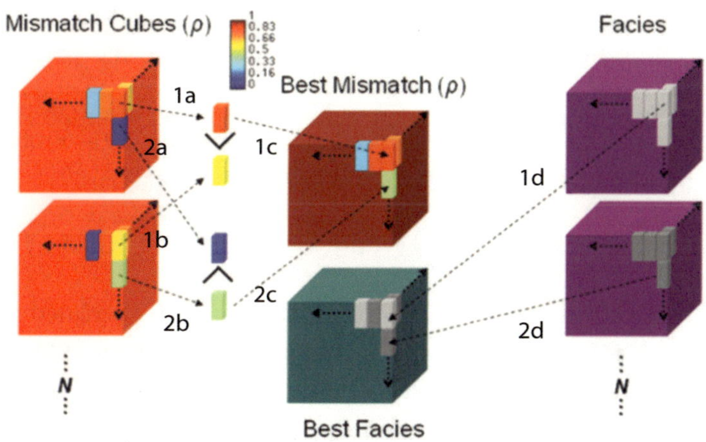

Figure 19. Process of creating "Best Mismatch cube" and "Best Facies cube".

As the methodology for acoustic impedance models, the process starts by analyzing each of the mismatch cubes that were created previously, and in each position of the cubes (x_0) it compares which has the higher value of correlation.

The point set 1a and point set 1b are compared, from those is the 1a that has a higher value of correlation so, it is copied (1c) to the Best Mismatch cube.

At the same time, since the first simulation showed better coefficient of correlation (CC) with the real seismic, another cube is created with the correspondent facies values (1d).

The process follow to the next set of values from the CC cubes, and it is the same, comparison between the sets 2a and 2b, that in this case it is the second cube that has the higher value of cc (or lower mismatch), it is these values that are copied to the Best Mismatch cube (2c). The same is done from the facies cube of this simulation, copy of the facies values to the Best Facies (2d). This process continues until all the N simulated models have been analyzed and to all the values of the cubes.

Through this way two new cubes are created, one with the best genes from each facies model (Best Facies) generated previously and another cube with the confidence factor of each part (Best Mismatch). As noticed above, the "least mismatch cube" can be seen as a summary of the least mismatch of all N realizations. The "best facies cube" can be seen as facies model combined from all N facies models that best matches the seismic data. However the "best facies cube" does not have the same geological concept as the training image and may have various artifacts, since it is constructed by copying several sets from independently generated facies models.

The third difference is that, these two new cubes can not be used for the co-simulation of the new generation of facies models, but the "best facies cube" can be used indirectly to improve the existing N facies models.

In order to do this, one uses a modification of the probability perturbation method as in [24]. Caers' method was developed to solve non-linear inverse problems under a prior model constraint. It allows the conditioning of stochastic simulations to any type of non-linear data. The principle of this method relies on perturbing the probability models used to generate a chain of realizations that converge to match any type of data.

In this methodology the unknown pre-posterior probability – Prob(A_j | C) – is modeled using a single parameter model in the following equation 9:

$$P\left(A_j \mid C\right) = P(I(u_j)=1 \mid C) = (1-r_c)\times i_B^{(0)}(u_j) + r_c \times P(A_j), j=1,\dots,N \tag{9}$$

where A is unknown data, B is well data and previously simulated facies indicators and C will be the "best facies cube", r_c is not dependent on u_j and is between [0,1], and $\left\{ i_B^{(0)}\left(\mathbf{u}_j\right),\ j=1,\dots,N \right\}$ is an initial realization conditioned to the B-data only.

According to the methodology proposed in this work, the equation 9 is adapted with r_c now representing the mismatch between the field and synthetic seismic as summarized with the least-mismatch cube (correlation coefficient), and $i_B^{(0)}$ corresponding to the presence of the facies with the least mismatch. This adaptation leads to the following equation 10:

$$P\left(A_j \mid C\right) = \rho_c(u) \times I(u_j) + (1 - \rho_c(u)) \times P(A_j), j = 1,...,N \tag{10}$$

where ρc is the correlation coefficient, between [0,1], extracted from the "least mismatch cube", $I(u_j)$ is the sand indicator extracted from the "best facies cube" with j=1,…,N facies, and $P(A_j)$ is the global proportion of the considered facies. This expression generates a "facies-probability cube" which is a mixture of global proportion and the "best facies cube".

This facies probabilities cube is then used as soft constraint to generate the next set of N cubes. This is done using Journel's tau-model to integrate probabilistic information from various sources (see equation 11 and figure 20);

$$\frac{x}{b} = \frac{c}{a} \tag{11}$$

where: $x = \dfrac{1 - P(A \mid B,C)}{P(A \mid B,C)}$, $b = \dfrac{1 - P(A \mid B)}{P(A \mid B)}$, $c = \dfrac{1 - P(A \mid C)}{P(A \mid C)}$ and $a = \dfrac{1 - P(A)}{P(A)}$

a is the information of the global facies proportion, *b* is the influence of the training image, and *c* is the conditioning of the soft probability cube.

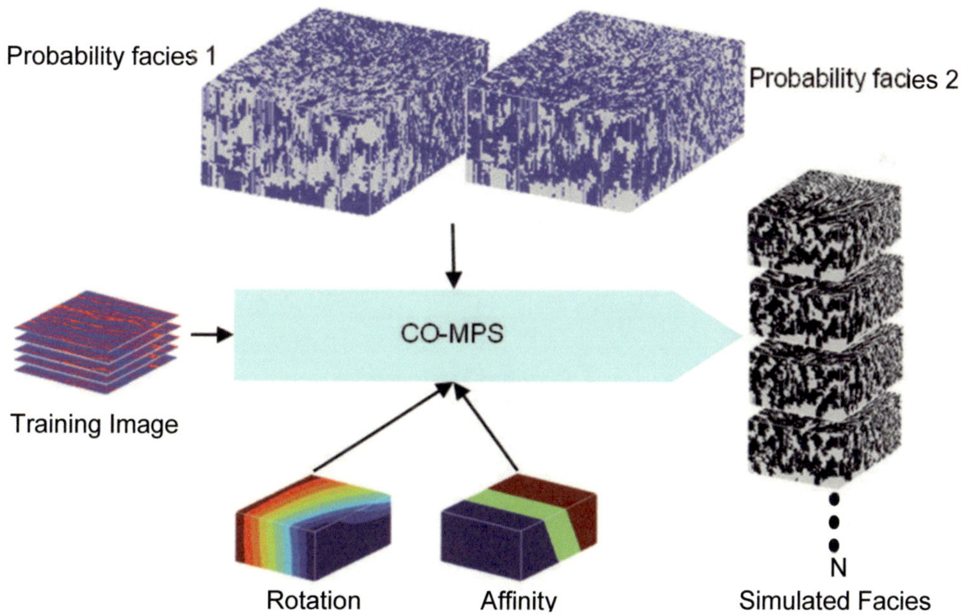

Figure 20. Next generation of simulations conditioned to the training image, rotation/affinity and probability cubes.

This iterative algorithm is run until the global mismatch between the synthetic seismic of the facies model and the field seismic data reach an optimal minimum.

5.5. Results

To illustrate the method, 6 iterations were computed with N=30 simulations on the Stanford VI dataset. Note that one assumed the availability of perfect geological information and a perfect forward model.

The algorithm converges as shown by a systematic increase in the global correlation coefficient between model and amplitude data (figure 21). The facies model with the highest correlation out of 30 models in the last iteration has a correlation of 0.88.

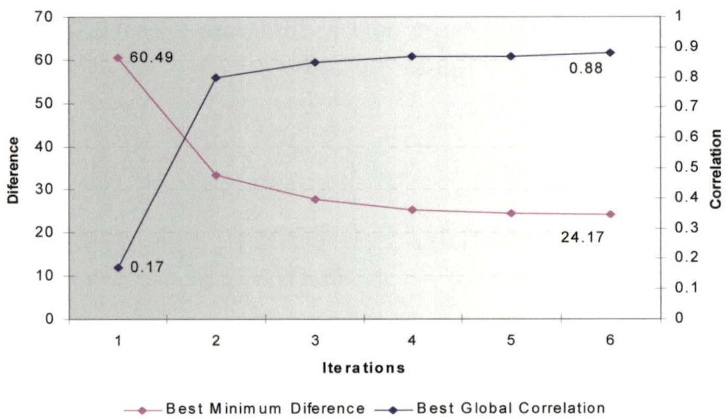

Figure 21. Convergence of the algorithm.

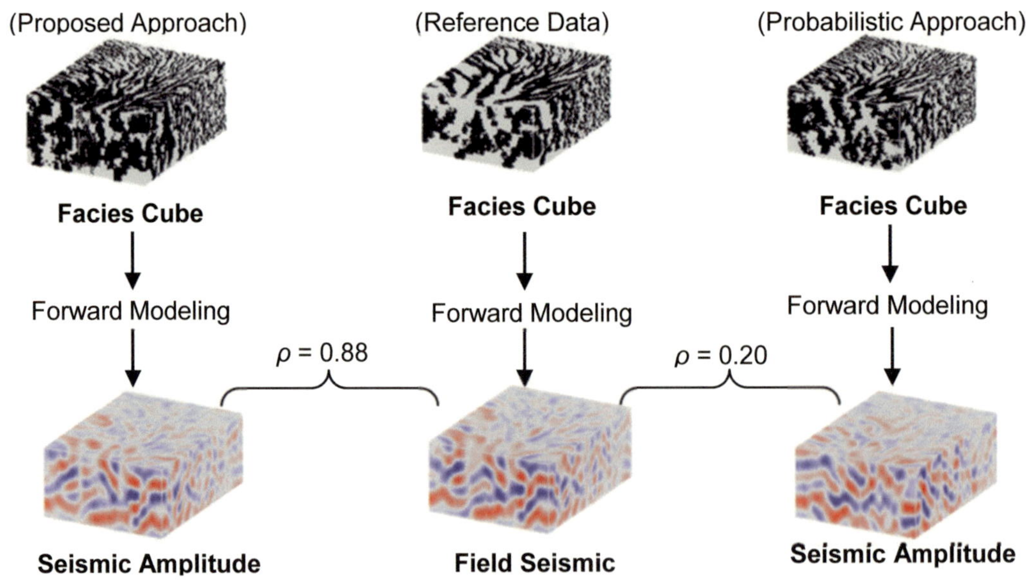

Figure 22. Comparison between the two methods

The final facies model (top left in in figure 22) does not contain any artifacts, i.e. reproduces well the geological continuity of the training image.

To check how well the seismic amplitude data is reproduced, is forward modeling the best facies model to seismic synthetic (see figure 22). Clearly the method matches well the field seismic, particularly when compared with the probabilistic approach.

In figure 23 some slices of the results are presented, where it is possible to see some similarities between the seismic response of the proposed approach and the reference data seismic amplitude.

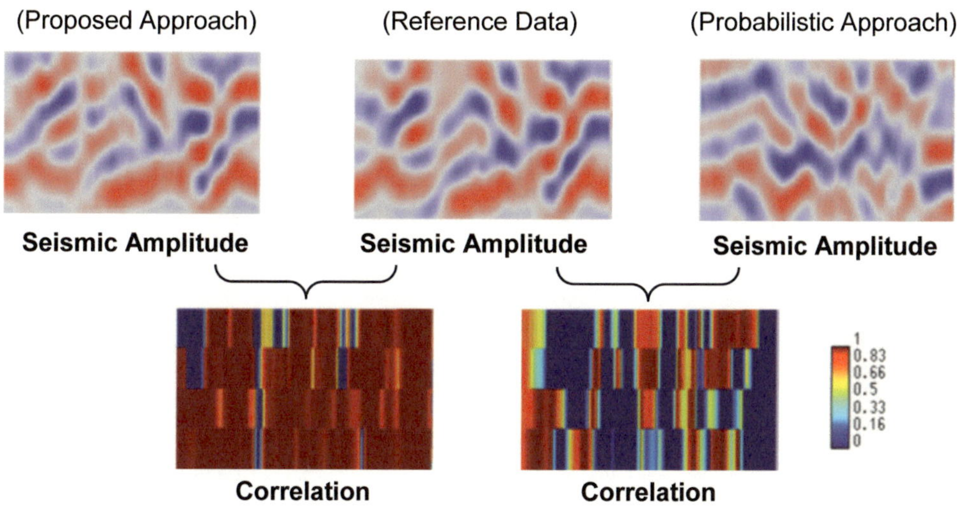

Figure 23. Correlation slices

In summary, as stated previously, the local maximum correlation or local minimum mismatch depends on the parameters used to calculate the mismatches. Some parameters sensitivity is further analyzed:

- Number of iterations

In the first iteration the main optimization is obtained, after this the process tends to stabilize (figure 24). Hence, without changing any of the other parameters, the number of iterations does not have a big influence on the optimization.

Figure 24 compares the values of correlation and average difference of the simulation model with the best result for different number of facies models N simulated. These values are presented only for the first and last iterations.

For the first iteration the difference is not very noticeable, but in the last iteration the correlation value is higher as the number of simulations increases.

- Number of facies simulations (N) per iterations

This parameter can have a major influence on the optimization: with numerous simulations the process has a wide variety of possibilities to choose from, and can build a more precise best-facies cube, but if the simulations are very alike, the choice of N will not matter.

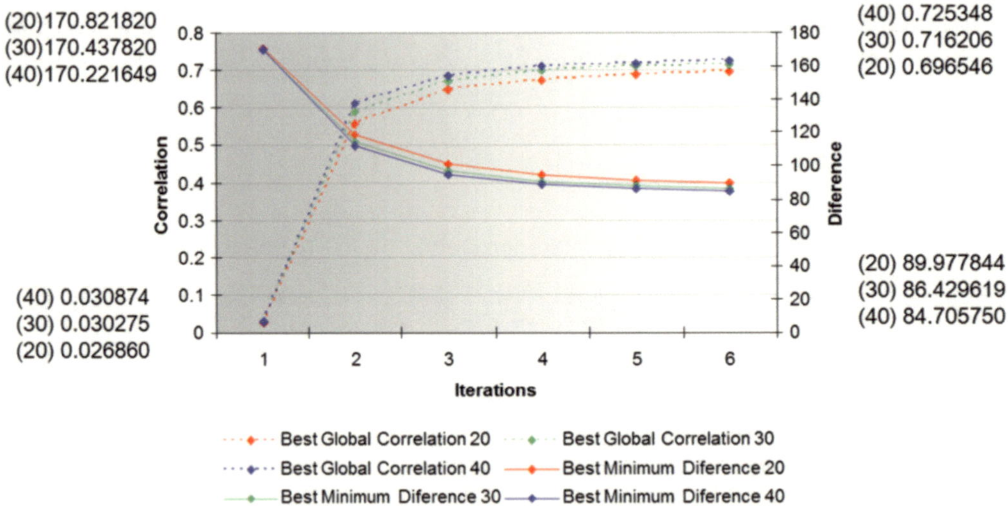

Figure 24. Influence of different number of iterations (1 to 6) and simulations (20, 30 and 40).

- Criterion of convergence

The criterion of differences is more precise than the correlation, since the correlation gives us a value of the similarity between two sets of data, while the difference is more susceptible to different patterns or small variation in the patterns. But to calculate the probabilities the correlations are needed.

- Size of the columns

In the previous test the cubes were divided in 4 layers with 20 values each column. Since the correlation has some sensitivity to the number of data used, the results also change. For a largest column the convergence will tend to be slower, since it is more difficult to match an entire column than piece-wise matching. In general we recommend that the number of layers chosen depends on the vertical resolution of the seismic data, i.e. for lower resolution less layer could be retained.

6. Conclusions

The algorithm proposed uses two different approaches to create a stochastic model of a reservoir property that are conditioned to both well and seismic data, since this two data are totally different, this process was to be an iterative one, that used the maximum correlation or minimum mismatch as the objective function.

The first approach uses the direct sequential co-simulation as the method of "transforming" 3D images, in an iterative process. Hence it generates, at each iterative step, global acoustic impedances images with the same spatial pattern, without imposing artificially good match in areas of low confidence in seismic quality.

It means that in those areas the final images will reflect high uncertainty. This uncertainty of the seismic acoustic impedance is used to access the uncertainty associated with the porosity model, by co-simulate the porosity model and using the generated final acoustic impedance model and its correlation coefficient cube as soft or secondary data.

Another example is the capacity to improve the seismic resolution, if the acoustic impedance model is created below the original resolution of the seismic. This is possible since the data from wells, that is mainly used to create the model, has a much lower resolution than the seismic.

Time consuming is no more a drawback as it is implemented in a parallel computing platform in [32].

This algorithm has been implemented and benchmarked tested with several real reservoirs, e.g. [33].

For the case of the multipoint approach, it is a new application for the newest stochastic generation of images and is a complement for the first approach.

The modifications made from the first approach to the second, are not too complex, except the population of the acoustic impendence cube, which for this example, a technique of simple populate the channels was used with previous simulated acoustic impedance values.

To resolve this bypass, an algorithm by [34], simulates any variable within the channel, making this step a more realistic one.

The program still needs to be improved and the algorithm can be optimized for real case applications and the lack of training images that could implement it to all kind or examples is still in research.

Author details

Hugo Caetano
Partex Oil and Gas, Portugal

Acknowledgement

This work was created based on the research made at: Centro de Modelização de Reservatórios Petrolíferos (CMRP), of Instituto Superior Técnico (IST), Universidade Técnica de Lisboa (UTL) with sponsorship from Partex - Oil and Gas and Stanford Center for

Reservoir Forecast (SCRF), Stanford University, Califórnia, EUA, with financial support from Fullbright Fundation and Fundação Luso-Americana.

7. References

[1] Soares, A., 1998, "Sequential Indicator with Correlation for Local Probabilities", Mathematical Geology, Vol 30, nº 6, p. 761-765

[2] Deutsch, C. V.; Journel, A. G., 1998, "GSLIB – Geostatistical Software Library and User's Guide", 2ª ed., Oxford University Press, New York

[3] Soares, A., 2001, "Direct Sequential Simulation and Co-Simulation", Mathematical Geology, Vol. 33, No. 8, p. 911-926

[4] Xu, W., Tran T.T., Srivastava, M., Journel A., 1992, "Integrating Seismic data in reservoir modeling: the collocated cokriging alternative". SPE #24742

[5] Doyen, P.M.; Psaila, D. E.; Strandenes, S., 1994, "Bayesian Sequential Indicator Simulation of Channel Sands from 3-D Seismic Data in The Oseberg Field, Norwegian North Sea", SPE 28382.

[6] Dubrule, O., 1993 "Introducing More Geology in Stochastic Reservoir Modeling," Geostatistics Troia 92, Soares Ed., Vol 1 pp 351-370

[7] Bortolli L., Alabert F., Haas A., Journel A.,1993. "Constraining Stochastic Images to Seismic Data". Geostatistics Troia 92, Soares Ed.,, Vol 1. pp 325-338

[8] Haas A., Dubrule O., 1994, "Geostatistical Inversion – A sequential Method for Stochastic Reservoir Modelling Constrained by Seismic Data" First Break (1994) 12, n-11, 561-569

[9] Grijalba-Cuenca A., Torrres-Verdin C., 2000, "Geostatistical Inversion of 3D Seismic Data to Extrapolate Wireline Petrophysical Variables Laterally away from the Well", SPE 63283

[10] Journel, A. G., 2002, "Combining Knowledge from Diverse Sources: an Alternative to Traditional Data Independence Hypothesis", Mathematical Geology, v. 34, No. 5, July 2002

[11] Mantilla, A., 2002, "Predicting Petrophysical Properties by Simultaneous Inversion of Seismic and Reservoir Engineering Data", PhD thesis, Stanford University

[12] Tarantola, A., 1984, "Inversion of Seismic Reflection Data in the Acoustic Approximation" Geophysics 49, 1259-1266

[13] Russell, B.H., 1988, "Introduction to Seismic Inversion Methods", Society of Exploration Geophysicists, 90 pp.

[14] Sheriff, R. E. and L. P. Geldart, 1995, Exploration Seismology, Cambridge University Press, New York, 592 pp.

[15] Caetano, H., Caers, J., 2006, "Constraining Multiple-point Facies Models to Seismic Amplitude Data", SCRF 2006

[16] Caetano, H., Caers, J., 2007, "Geostatistical Inversion Constrained to Multi-Point Statistical Information Derived from Training Images", EAGE – Petroleum Geostatistics, Cascais, September 2007

[17] Soares, A., Diet, J.D., Guerreiro, L., 2007, "Stochastic Inversion with a Global Perturbation Method", EAGE – Petroleum Geostatistics, Cascais, September 2007

[18] Carvalho, J., Soares, A., Bio, A., 2006, "Classified Satellite Sensor Data Calibration with Geostatistical Stochastic Simulation", International Journal of Remote Sensing, 27(16), p. 3375-3386.

[19] Mata-Lima, H., 2008, "Reservoir Characterization with Iteractive Direct Sequencial Co-Simulation: Integrating Fluid Dynamic Data into Stochastic Model", Journal of Petroleum Science and Engineering, 62(3-4), p. 59-72.

[20] Strebelle, S., 2002, "Conditioning Simulation of Complex Structures Multiple-Point Statistics", Mathematical Geology, v. 34, No. 1, January 2002.

[21] Strebelle, S., 2001, "Sequencial Simulation drawing structures from training images", PhD thesis, Stanford University

[22] Hulstrand R. F., Abou Choucha, M. K. A., Al Baker, S. M., 1985, "Geological Model of the Bu hasa / Shuaiba Reservoir, Society of Petroleum Engineers

[23] Guerreiro, L., Caetano, H., Maciel, C., Real, A., Silva, F., Al Shaikh, A., Al Shemsi, M. A., 2007, "Geostatistical Seismic Inversion Applied to a Carbonate Reservoir", EAGE – Petroleum Geostatistics, Cascais, September 2007

[24] Caeiro H., 2007, "Caracterização da Incerteza da Porosidade de um Reservatório Petrolífero", MSc Thesis, Instituto Superior Técnico

[25] Stoyan, D., Kendall, W., Mecke, L., 1987, "Stochastic geometry and its applications" Wiley, N.Y.

[26] Haldorsen, H., Damsleth, E., 1990, "Stochasting Modeling", Journal or Petroleum Technology, April 1990, p 404-412.

[27] Srivastava, R. M,, 1992, "Reservoir Characterization with Probability field Simulation", SPE paper no. 24753

[28] Caers, J., Journel, A. G., 1998, "Stochasting Reservoir Simulation Using Neural Networks Trained on Outcrop Data", SPE paper no. 49026

[29] Castro, S., Caers, J and Mukerji Y, 2005, "The Stanford VI reservoir", SCRF 2005, Stanford University

[30] Caumon, G., Journel, A., 2005, "Early Uncertainty Assessment: Application to a Hydrocarbon Reservoir Appraisal", Quantitative Geology and Geostatistics Volume 14, Pages 551-557

[31] Caers, J., 2003, "History matching under training-image based geological model constraints", SPE Journal, v. 7: 218-226

[32] Vargas, H., Caetano, H., Filipe, M., 2007, "Parallelization of Sequential Simulation Procedures", EAGE – Petroleum Geostatistics, Cascais, September 2007

[33] Guerreiro, L., Caetano, H., Maciel, C., Real, A., Silva, F., Al Shaikh, A., Ali Al Shemsi M. A., Soares A., 2007, "Global Seismic Inversion - A New Approach to Integrate Seismic

Information in the Stochastic Models", SPE 111305, 2007 SPE/EAGE Reservoir Characterization and Simulation Conference held in Abu Dhabi

[34] Horta A., Caeiro M.H., Nunes R., Soares A, 2008, "Simulation of Continuous Variables at Meander Structures: Application to Contaminated Sediments of a Lagoon"

Section 3

Reservoir Management

Chapter 7

Transient Pressure and Pressure Derivative Analysis for Non-Newtonian Fluids

Freddy Humberto Escobar

Additional information is available at the end of the chapter

http://dx.doi.org/10.5772/50415

1. Introduction

Conventional well test interpretation models do not work in reservoirs containing non-Newtonian fluids such as completion and stimulation treatment fluids: polymer solutions, foams, drilling muds (this should not be considered as a reservoir fluid, since before testing we should clean the well to remove all the drilling invasion fluids, however it obeys the power-law), etc., and some paraffinic oils and **heavy crude oils**. Non-Newtonian fluids are generally classified as time independent, time dependent and viscoelastic. Examples of the first classification are the Bingham, pseudoplastic and dilatant fluids, Figure 1, which are commonly dealt by petroleum engineers.

As a special kind of non-Newtonian fluid, Bingham fluids (or plastics) exhibit a finite yield stress at zero shear rates. There is no gross movement of fluids until the yield stress, τ_y, is exceeded. Once this is accomplished, it is also required cutting efforts to increase the shear rate, i.e. they behave as Newtonian fluids. These fluids behave as a straight line crossing the y axis in $\tau = \tau_y$, when the shear stress, τ plotted against the shear rate, γ in Cartesian coordinates. The characteristics of these fluids are defined by two constants: the yield, τ_y, which is the stress that must be exceeded for flow to begin, and the Bingham plastic coefficient, μ_B. The rheological equation for a Bingham plastic is,

$$\tau = \tau_y + \mu_B \gamma$$

The Bingham plastic concept has been found to approximate closely many real fluids existing in porous media, such as paraffinic oils, heavy oils, drilling muds and fracturing fluids, which are suspensions of finely divided solids in liquids. Laboratory investigations have indicated that the flow of heavy-oil in some fields has non-Newtonian behavior and approaches the Bingham type.

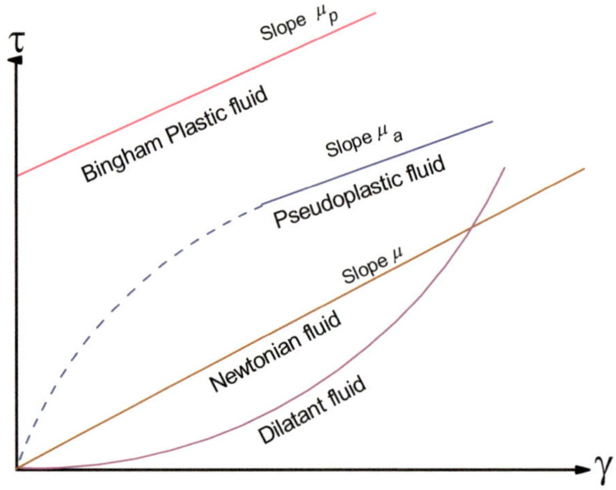

Figure 1. Schematic representation of time-independent fluid

Pseudoplastic and dilatant fluids have no yield point. The slope of shear stress versus shear rate decreases progressively and tends to become constant for high values of shear stress for pseudoplastic fluids. The simplest model is power law,

$$\tau = k\gamma^{n}; \quad n < 1.$$

k and n are constants which differ for each particular fluid. k measures the flow consistency and n measures the deviation from the Newtonian behavior which $k = \mu$ and $n = 1$.

Dilatants fluids are similar to pseudoplastic except that the apparent viscosity increases as the shear stress increases. The power-law model also describes the behavior of dilatant fluids but $n > 1$.

Currently, unconventional reservoirs are the most impacting subject in the oil industry. Shale reservoirs, coalbed gas, tight gas, gas hydrates, gas storage, geothermal energy, coal – conversion to gas, coal-to-gas, in-situ gasification and heavy oil are considered unconventional reservoirs. In the field of well testing, several analytical and numerical models taking into account Bingham, pseudoplastic and dilatant non-Newtonian behavior have been introduced in the literature to study their transient nature in porous media for a better reservoir characterization. Most of them deal with fracture wells, homogeneous and double-porosity formations and well test interpretation is conducted via the straight-line conventional analysis or type-curve matching and recently some studies involving the pressure derivative have also been introduced.

When it is required to conduct a treatment with a non-Newtonian fluid in an oil-bearing formation, this comes in contact with conventional oil which possesses a Newtonian nature. This implies the definition of two media with entirely different mobilities. If a pressure test is run in such a system, the interpretation of data from such a test through the use of conventional straight-line method may be erroneous and may not provide a way for verification of the results obtained.

The purpose of this chapter is to provide the most updated state-of-the-art on transient analysis of Non-Newtonian fluids and to present both conventional and modern methodologies for well test interpretation in reservoirs saturated with such fluids. Especial interest is given to the use of the pressure and pressure derivative for both homogeneous and double-porosity formations.

2. Non-Newtonian fluids in transient pressure analysis

Non-Newtonian fluids are often used during various drilling, workover and enhanced oil recovery processes. Most of the fracturing fluids injected into reservoir-bearing formations behave non-Newtonianly and these fluids are often approximated by Newtonian fluid flow models. In the field of well testing, several analytical and numerical models taking into account Bingham and pseudoplastic non-Newtonian behavior have been introduced in the literature to study the transient nature of these fluids in porous media for a better reservoir characterization. Most of them deal with fracture wells and homogeneous formations and well test interpretation is conducted via the straight-line conventional analysis or type-curve matching. Only a few studies consider pressure derivative analysis. However, there exists a need of a more practical and accurate way of characterizing such systems.

Many studies in petroleum and chemical engineering and rheology have focused on non-Newtonian fluid behavior though porous formations, among them, we can name [6, 9, 10, 18, 20, 23]. Several numerical and analytical models have been proposed to study the transient behavior of non-Newtonian fluid in porous media. Since all of them were published before the eighties, when the pressure derivative concept was inexistent; interpretation technique was conducted using either conventional analysis or type-curve matching.

It is worth to recognize that Ikoku has been the researcher who has contributed the most to non-Newtonian power-law fluids modeling, as it is demonstrated in the works of [9,10,11,13]. All of these models have been used later for other researchers for further development of test interpretation techniques. For instance, reference [24] presented a study of a pressure fall-off behavior after the injection of a non-Newtonian power-law fluid. [14] presented a study using the elliptical flow on transient analysis interpretation in Polymer flooding EOR since polymer solutions also exhibit non-Newtonian rheological behavior such as in-situ shear-thinning and shear-thickening effects.

[25] used for the first time the pressure-derivative concept for well test analysis of non-Newtonian fluids, and later on, [12] presented the first extension of the *TDS* (Tiab's Direct Synthesis) technique, [21] to non-Newtonian fluids. [7] used type-curve matching for interpretation of pressure test for non-Newtonian fluids in infinite systems with skin and wellbore storage effects. Recent applications of the derivative function to non-Newtonian system solutions are presented by [1] and [15] who applied the *TDS* technique to radial composite reservoirs with a Non-Newtonian/Newtonian interface for pseudoplastic and dilatants systems, respectively.

As far as non-Newtonian fluid flow through naturally fractured reservoirs is concerned only a study presented by [19] is reported in the literature. He presented the analytical solution for the transient behavior of double-porosity infinite formations which bear a non-Newtonian pseudoplastic fluid and his analytical solution also considers wellbore storage effects and skin factor; therefore, [2] used the analytical solution without wellbore storage and skin introduced by [19] was used to develop an interpretation technique using the pressure and pressure derivative, so expressions to estimate the Warren and Root parameters [26] (dimensionless *storage coefficient and interporosity flow parameter)* were found and successfully tested with synthetic data.

3. Pseudoplastic infinite-acting radial flow regime in homogeneous formations

Interpretation of pressure tests for non-Newtonian fluids is performed differently to conventional Newtonian fluids. During radial flow regime, Non-Newtonian fluids exhibit a pressure derivative curve which is not horizontal but rather inclined. As shown by [12], the smaller the value of n (flow behavior index) the more inclined is the infinite-acting pressure derivative line, see Figure 2.

A partial differential equation for radial flow of non-Newtonian fluids that follow a power-law relationship through porous media was proposed [11]. Coupling the non-Newtonian Darcy's law with the continuity equation, they derived a rigorous partial differential equation:

$$\frac{\partial^2 P}{\partial r^2}+\frac{n}{r}\frac{\partial P}{\partial r}=c_t\phi n\left(\frac{\mu_{eff}}{k}\right)^{1/n}\left(-\frac{\partial P}{\partial r}\right)^{(n-1)/n}\frac{\partial P}{\partial t} \tag{1}$$

This equation is nonlinear. For analytical solutions, a linearized approximation was also derived by [11]:

$$\frac{1}{r^n}\frac{\partial}{\partial r}\left(r^n\frac{\partial P}{\partial r}\right)=Gr^{1-n}\frac{\partial P}{\partial t} \tag{2}$$

Where:

$$G=\frac{3792.188 n\phi c_t\mu_{eff}}{k}\left(96681.605\frac{h}{qB}\right)^{1-n} \tag{3}$$

and,

$$\mu_{eff}=\left(\frac{H}{12}\right)\left(9+\frac{3}{n}\right)^n\left(1.59344\times 10^{-12}k\phi\right)^{(1-n)/2} \tag{4}$$

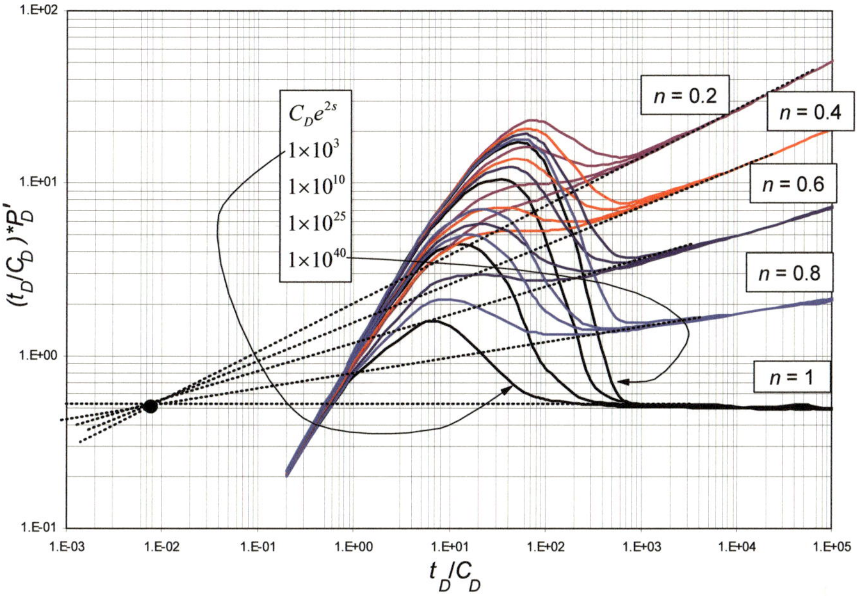

Figure 2. Pressure derivative for a pseudoplastic non-Newtonian fluid in an infinite reservoir – After Reference [12]

The dimensionless quantities were also introduced by [10] as

$$P_{DNN} = \frac{\Delta P}{141.2(96681.605)^{1-n}\left(\frac{qB}{h}\right)^n \frac{\mu_{eff} r_w^{1-n}}{k}} \tag{5}$$

$$t_{DNN} = \frac{t}{G r_w^{3-n}} \tag{6}$$

$$P_{DN} = \frac{k\ h\Delta P}{141.2 q \mu_N B} \tag{7}$$

$$t_{DN} = \frac{0.0002637k\ t}{\phi \mu_N c_t r_w^2} \tag{8}$$

$$r_D = \frac{r}{r_w} \tag{9}$$

Where suffix *N* indicates Newtonian and suffix *NN* indicates non-Newtonian. The dimensionless well pressure analytical solution in the Laplace space domain for the case of a well producing a pseudoplastic non-Newtonian fluid at a constant rate from an infinite reservoir is given in reference [11]:

$$\bar{P}_D(\tilde{s}) = \frac{K_\nu\left(\beta\sqrt{\tilde{s}}r_D^{1/\beta}\right) + s\sqrt{\tilde{s}}K_\beta\left(\beta\sqrt{\tilde{s}}\right)}{\tilde{s}\left(\sqrt{\tilde{s}}K_\beta\left(\beta\sqrt{\tilde{s}}\right) + \tilde{s}C_D\left[K_\nu\left(\beta\sqrt{\tilde{s}}\right) + s\sqrt{\tilde{s}}K_\beta\left(\beta\sqrt{\tilde{s}}\right)\right]\right)} \tag{10}$$

Being $\beta = 2/(3\text{-}n)$ and $\nu = (1\text{-}n)/(3\text{-}n)$.

The dimensionless pressure derivative during radial flow regime is governed by:

$$\left(t_D * P_D'\right)_{rNN} = 0.5t_{DNN}^{\alpha} \tag{11}$$

[12] presented the following expression to estimate the permeability,

$$\frac{k}{\mu_{eff}} = \left[0.5\frac{t_r^{\alpha}}{C^{\alpha}\left(t * \Delta P'\right)_r}\frac{\left(2\pi h\right)^{n(\alpha-1)}r_w^{(1-n)(1-\alpha)}}{q^{n(\alpha-1)-\alpha}}\right]^{\frac{1}{1-\alpha}} \tag{12}$$

where $\alpha = -0.1486n^2 - 0.178n + 0.3279$

being n the flow behavior index which may be found from the slope of the pressure derivative curve during radial flow regime. [12] also introduced another expressions and correlations to find permeability, skin factor and wellbore storage coefficient using the maximum point (peak) found on the pressure derivative curve during wellbore storage effects which are not shown here. The point of intercept between the early unit-slope line and radial flow regime is used to estimate wellbore storage:

$$t_i = \frac{\left(3.13e^{-1.85n}\right)C}{\left(2\pi h\right)^n}\frac{\mu_{eff}}{k}\left(\frac{q}{r_w}\right)^{n-1} \tag{13}$$

Parameters in both Equations 11 and 12 are given in CGS (centimeters, grams, seconds) units.

[1] presented more practical expressions for the determination of both permeability and skin factor:

$$\frac{k}{\mu_{eff}} = \left[70.6\left(96681.605\right)^{(1-\alpha)(1-n)}\left(\frac{0.0002637t_r}{n\phi c_t}\right)^{\alpha}\left(\frac{qB}{h}\right)^{n-\alpha(n-1)}\left(\frac{1}{\left(t * \Delta P'\right)_r}\right)\right]^{\frac{1}{1-\alpha}} \tag{14}$$

Where α is the slope of the pressure derivative curve and is defined by:

$$\alpha = \frac{1-n}{3-n} \tag{15}$$

$$s_{rNN} = \frac{1}{2}\left(\frac{(\Delta P)_{rNN}}{(t * \Delta P')_{rNN}} - \frac{1}{a} \right)\left(\frac{t_{rNN}}{G\, r_w^{3-n}} \right)^a \tag{16}$$

4. Well pressure behavior in non-Newtonian/Newtonian interface

In many activities of the oil industry, engineers have to deal with completion and stimulation treatment fluids such as polymer solutions and some heavy crude oils which obey a non-Newtonian power-law behavior. When it is required to conduct a treatment with a non-Newtonian fluid in an oil-bearing formation, this comes in contact with conventional oil which possesses a Newtonian nature. This implies the definition of two media with entirely different mobilities. If a pressure test is run in such a system, the interpretation of data from such a test through the use of conventional straight-line method may be erroneous and may not provide a way for verification of the results obtained. Then, [13] proposed a solution for the system sketched in Figure 3 which was solved numerically by [17].

[15] presented for the first time the pressure derivative behavior for the mentioned system, Figure 4. Notice in that plot that the pressure derivative shows an increasing slope as the flow behavior index decreases. Also, the derivative has no slope during infinite-acting Newtonian behavior, as expected.

During the non-Newtonian region, region 1 in Figure 3, Equations 13 to 15 work well. For the Newtonian region, region 2, the permeability and skin factor are estimated with the equations presented by Tiab (1993) as:

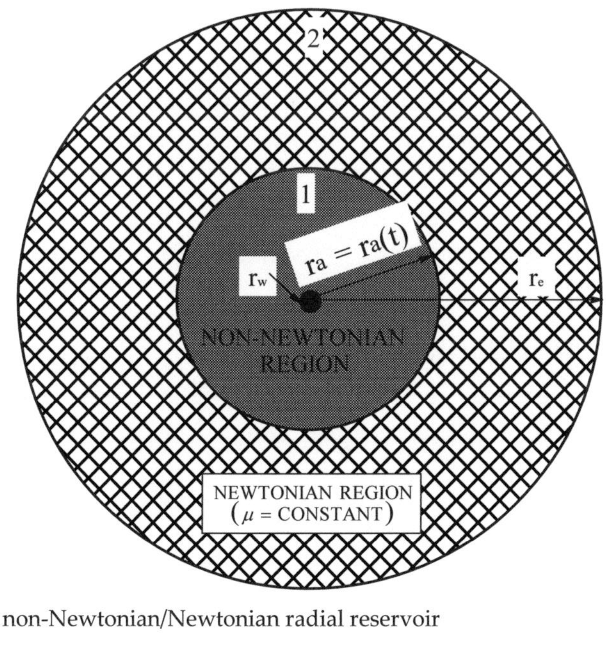

Figure 3. Composite non-Newtonian/Newtonian radial reservoir

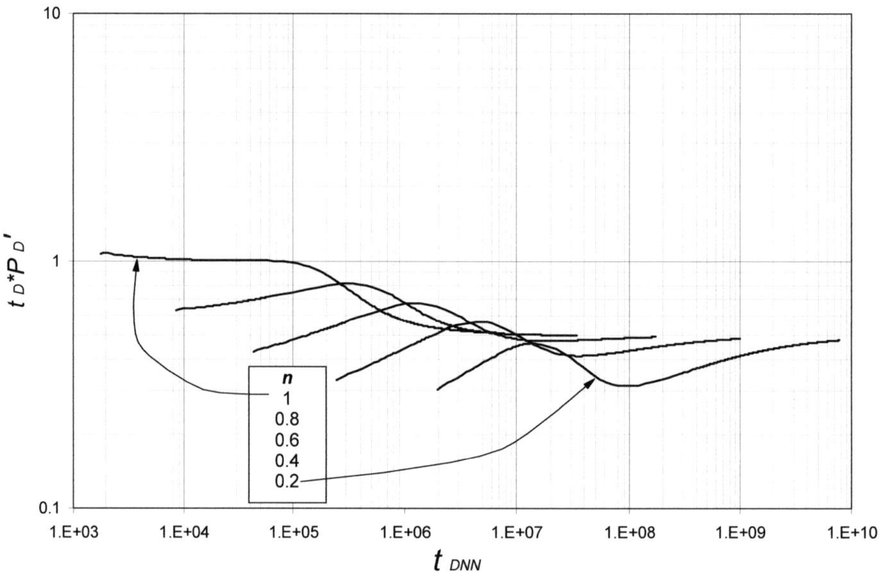

Figure 4. Dimensionless pressure derivative behavior for r_a = 200 ft. Case Non-Newtonian pseudoplastic

$$k_2 = \frac{70.6 q \mu_N B}{h(t*\Delta P')_r} \tag{17}$$

$$s_2 = \frac{1}{2}\left[\left(\frac{\Delta P}{t*\Delta P'}\right)_{r2} - \ln\left(\frac{k_2 t_{r2}}{\phi \mu_N c_t r_w^2}\right) + 7.43\right] \tag{18}$$

Suffix 2 denotes the non-Newtonian region.

[15] also found an expression to estimate the non-Newtonian permeability using the time of intersection of the non-Newtonian and Newtonian radial lines, t_{iN_NN}:

$$k = \left[\left\langle\left(\frac{H}{12}\right)\left(9+\frac{3}{n}\right)^n \left(1.59344\times10^{-12}\phi\right)^{(1-n)/2}\left(96681.605\frac{hr_w}{qB}\right)^{1-n}\right\rangle^{1-1/\alpha} \frac{\mu_N^{1/\alpha}\phi c_t r_w^2 n}{0.0002637 t_{irN\_NN}}\right]^{1/2} \tag{19}$$

The radius of the injected non-Newtonian fluid bank is calculated using the following correlation (not valid for n=1), obtained from reading the time at which the pressure derivative has its maximum value:

$$r_a = \left[\frac{G\left(0.2258731n^3 - 0.2734284n^2 + 0.5064602n + 0.5178275\right)^{1/\alpha}}{t_{MAX}}\right]^{1/(n-3)} \tag{20}$$

[13] found that the radius of the non-Newtonian fluid bank can be found using the radius investigation equation proposed by [10]:

$$r_a = \left[\Gamma\left(\frac{2}{3-n}\right)\right]^{1/(n-1)} \left[\frac{(3-n)^2 t}{G}\right]^{1/(3-n)} \tag{21}$$

where t is the end time of the straight line found on a non-Newtonian Cartesian graph of ΔP vs. $t^{1-n/3-n}$.

Later, [16] found that Equations 13, 14, 15 and 22 also worked for dilatant systems. This is the case when $2 < n < 1$. The pressure derivative behavior is given in Figure 5. Notice that for this case the slope decreases as the flow behavior index increases. For dilatant-Newtonian interface the position of the front obeys the following equation:

$$r_a = \left[\frac{G\left(0.46811e^{0.76241n}\right)^{1/\alpha}}{t_{e\_rNN}}\right]^{1/(n-3)} \tag{22}$$

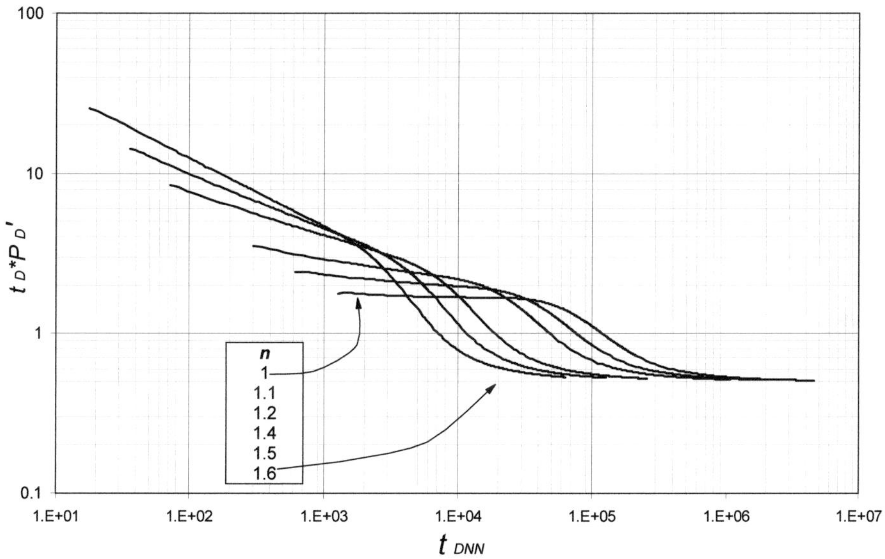

Figure 5. Dimensionless pressure derivative behavior for r_a = 200 ft. Case Non-Newtonian dilatant

Example 1. A constant-rate injection test for a well in a closed reservoir was generated by [13] with the data given below. It is required to estimate the permeability and the skin factor in each area and the radius of injected non-Newtonian fluid bank.

P_R = 2500 psi	r_e = 2625 ft	r_w = 0.33 ft	h = 16.4 ft
ϕ = 20 %	k = 100 md	q = 300 BPD	B = 1.0 rb/STB
c_t = 6.89x10^{-6} 1/psi	r_a = 131.2 ft	H = 20 cp*s^{n-1}	μ_N = 3 cp

n = 0.6

Solution. The log-log plot of pressure and pressure derivative against injection time is given in Figure 6. Suffix 1 and 2 indicate the non-Newtonian and Newtonian regions, respectively. From Figure 6 the following information was read:

$t_{r1} = 0.3$	$\Delta P_{r1} = 541.54$ psi	$(t^*\Delta P')_{r1} = 105.45$ psi
$t_{r2} = 120$	$\Delta P_{r2} = 991.5$ psi	$(t^*\Delta P')_{r2} = 39.02$ psi
$t_{MAX} = 1.3$	$t_{irN\_NN} = 0.0008$ hr	

First, α is evaluated with Equation 15 to be 0.17 and a value of 100.4 md was found with Equation 14 for the non-Newtonian effective fluid permeability. Equation 4 is used to find an effective viscosity of 0.06465 cp(s/ft)$^{n-1}$. Then, the skin factor in the non-Newtonian region is found with Equation 16 to be 179.7.

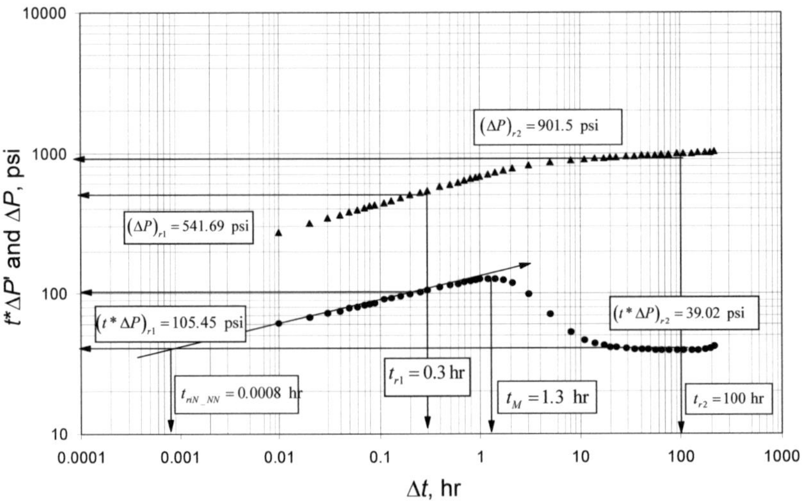

Figure 6. Pressure and pressure derivative for example 1

A value of 6.228x10^{-5} hr/(ft^{3-n}) was found for parameter G using Equation 3. This value is used in Equation 24 to find the distance from the well to the non-Newtonian fluid bank. This resulted to be 120 ft.

Equations 17 and 18 were used to estimate permeability and skin factor of the Newtonian zone. They resulted to be 100 md and 4.5.

Using a time of 0.0008 hr which corresponds to the intersect point formed between the non-Newtonian and Newtonian radial flow regime lines in Equation 19, a non-Newtonian effective fluid permeability of 96 md is re-estimated. [13] obtained a permeability of the non-Newtonian zone of 101 md and $r_a = 116$ ft from conventional analysis.

5. Hydraulically fractured wells

[18] linearized the partial-differential equation for the problem of a well intercepted by a vertical fracture. Their dimensionless pressure solution is given below:

$$P_D\left(t_D\right)=\frac{(3-n)^{2v}t_D^{v}}{\left(1-n\right)\Gamma(1-\nu)}-\frac{1}{1-n} \tag{23}$$

Where $\nu=(1-n)/(3-n)$

[24] presented two interpretation methodologies: type-curve matching and conventional straight-line for characterization of fall-off tests in vertically hydraulic wells with a pseudoplastic fluid. They indicated that at early times, a well-defined straight line with slope equal to 0.5 on log-log coordinates will be evident, then,

$$P_D=\left(\frac{\pi}{2}\right)^{\frac{n-1}{2}}\sqrt{\pi * t_{Dxf}} \tag{24}$$

$$t_{Dxf}=\frac{0.0002637kt}{\phi c_t\mu^* x_f^2} \tag{25}$$

Where the characteristic viscosity, μ^*, is given by:

$$\mu^*=\mu_{eff}\left(96681.605\frac{h}{qB}\right)^{1-n} \tag{26}$$

And the derivative of Equation 24 is:

$$t_{Dxf} * P_D{}'=0.5\left(\frac{\pi}{2}\right)^{\frac{n-1}{2}}\sqrt{\pi t_{Dxf}} \tag{27}$$

And the dimensionless fractured conductivity is;

$$c_{fD}=\frac{k_f w_f}{k\,x_f} \tag{28}$$

[22] presented an expression which relate the half fracture length, x_f, formation permeability, k, fracture conductivity, $k_f w_f$, and post-frac skin factor, s:

$$k_f w_f=\frac{3.31739k}{\frac{e^s}{r_w}-\frac{1.92173}{x_f}} \tag{29}$$

However, there is no proof that Equation 28 works for Non-Newtonian systems. Using Equation 23, [3] presented pressure and pressure derivative curves for vertically infinite-conductivity fractured wells. See Figure 7. They extended the *TDS* methodology, [21], for the systems under consideration. By using the intersect point of the pressure derivatives during linear flow regime, Equation 26, with the radial flow regime governing equation, Equation 11, t_{RLi}, an expression to obtain the half-fracture length is presented:

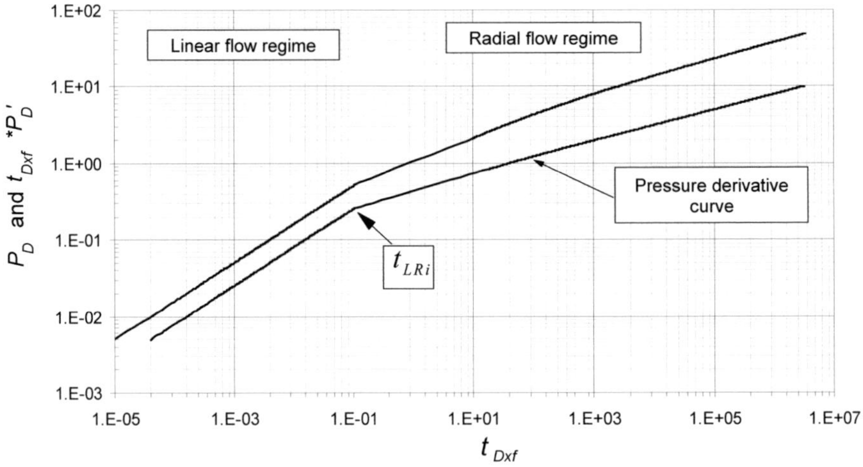

Figure 7. Dimensionless pressure and pressure derivative behavior for a vertical infinite-conductivity fractured well with a non-Newtonian pseudoplastic fluid with $n = 0.5$

$$x_f = \left[0.028783 \frac{(1.570796)^{\frac{n-1}{2}}}{\left(\frac{0.0002637 k t_{LRi}}{\phi c_t \mu^*} \right)^{\alpha}} \sqrt{\frac{t_{LRi} k}{\phi c_t \mu^*}} \right]^{\alpha} \tag{30}$$

Where $\nu = (1\text{-}n)/(3\text{-}n)$.

The expression governing the late-time pseudosteady-state flow regime is:

$$t_D * P_D' = 2\pi t_{DA} \tag{31}$$

The point of intersection of the pressure derivatives during linear flow and pseudosteady-state (mathematical development is not shown here) allows to obtain the well drainage area by means of the following expression:

$$A = \pi \left[\frac{t_{iLPSS}}{0.0625 \left(\frac{\pi}{2} \right)^{n-1} G} \right]^{2/(3-n)} \tag{32}$$

Example 2. Fan (1998) presented a pressure test of a test conducted in a hydraulic fractured well with the information given below. Pressure and pressure derivative data for this test is reported in Figure 8.

$n = 0.4$	$h = 70$ ft	$k = 0.65$ md	$q = 507.5$ BPD
$\phi = 10$ %	$B = 1$ rb/STB	$\mu^* = 0.00065$ cp	$c_t = 0.00001$ psi^{-1}

$r_w = 0.26$ ft $\qquad H = 20$ cp*s^{n-1}

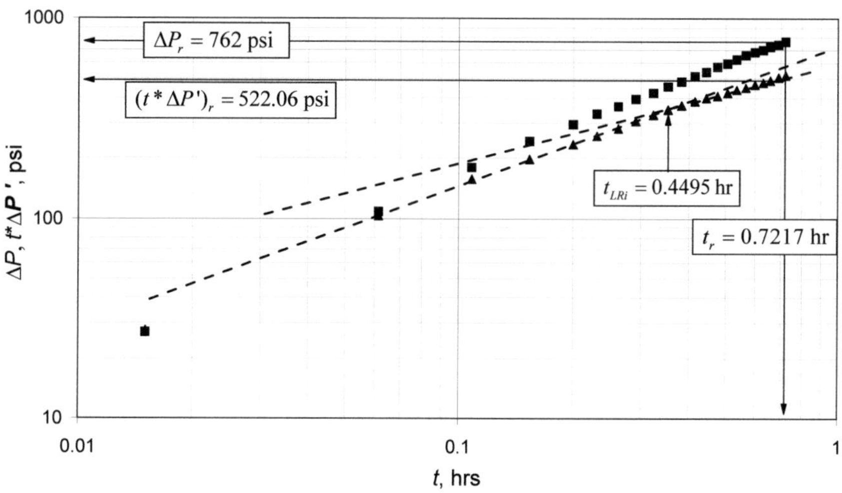

Figure 8. Pressure and pressure derivative for example 2

Solution. The following information was read from the pressure and pressure derivative plot, Figure 8,

t_{LRi}=0.4495 hr $\qquad t_r$=0.7217 hr $\qquad \Delta P_r$= 762 psi $\qquad (t*\Delta P')_r$ = 522.06 psi

Using Equation 15, a value of 0.23 is found for α. Reservoir permeability, skin factor, half-fracture length were estimated with Equations 14, 16 y 30. Their respective values are 0.65 md, -13.9 and 771 ft. Reservoir permeability and half-fracture length are re-estimate by simulating the test providing values of 0.65 md and 776 ft, respectively; therefore, the absolute errors for these calculations are less 0.06 % and 0.5 %. A G value of 0.001241 hr/(ft^{3-n}) was found with Equation 3.

A fracture conductivity of 868.5 md-ft was calculated using Equation 29. It is important to clarify that this equation is valid for the Newtonian case. This value was used in Equation 28 to find a dimensionless fracture conductivity of 1.73.

6. Finite-homogeneous reservoirs

For the cases of bounded and constant-pressure reservoirs, [8] presented the solutions to Equation 1. The initial and boundary conditions for the first case are:

$$P_{DNN}(r_D, 0) = 0 \tag{33}$$

$$\left(\frac{\partial P_{DNN}}{\partial r_D}\right)_{r_D=1} = -1 \;\; for\; t_{DNN} > 0 \tag{34}$$

$$\left(\frac{\partial P_{DNN}}{\partial r_D}\right)_{r_{eD}=1} = 0 \quad for \; t_{DNN} \tag{35}$$

The analytical solution in the Laplace space domain for the closed reservoirs under constant-rate case is given as:

$$\overline{P}(\tilde{s}) = \frac{\left\{K_{2/(3-n)}\left[\frac{2}{3-n}\sqrt{\tilde{s}}r_{eD}^{(3-n)/2}\right] \bullet I_{\frac{1-n}{3-n}}\left[\frac{2}{3-n}\sqrt{\tilde{s}}\right] + I_{2/(3-n)}\left[\frac{2}{3-n}\sqrt{\tilde{s}}r_{eD}^{(3-n)/2}\right] \bullet K_{\frac{1-n}{3-n}}\left[\frac{2}{3-n}\sqrt{\tilde{s}}\right]\right\}}{\left(\tilde{s}^{3/2}\left\{I_{2/(3-n)}\left[\frac{2}{3-n}\sqrt{\tilde{s}}r_{eD}^{(3-n)/2}\right] \bullet K_{2/(3-n)}\left[\frac{2}{3-n}\sqrt{\tilde{s}}\right] - K_{2/(3-n)}\left[\frac{2}{3-n}\sqrt{\tilde{s}}r_{eD}^{(3-n)/2}\right] \bullet I_{2/(3-n)}\left[\frac{2}{3-n}\sqrt{\tilde{s}}\right]\right\}\right)} \tag{36}$$

For the case of constant-pressure external boundary, the boundary condition given by Equation 35 is changed to:

$$P_{DNN}\left(r_{eD}, t_{DNN}\right) = 0 \tag{37}$$

And the analytical solution for such case is:

$$\overline{P}(\tilde{s}) = \frac{\left\{I_{\frac{1-n}{3-n}}\left[\frac{2}{3-n}\sqrt{\tilde{s}}r_{eD}^{(3-n)/2}\right] \bullet K_{\frac{1-n}{3-n}}\left[\frac{2}{3-n}\sqrt{\tilde{s}}\right] - K_{\frac{1-n}{3-n}}\left[\frac{2}{3-n}\sqrt{\tilde{s}}r_{eD}^{(3-n)/2}\right] \bullet I_{\frac{1-n}{3-n}}\left[\frac{2}{3-n}\sqrt{\tilde{s}}\right]\right\}}{\left(\tilde{s}^{3/2}\left\{I_{2/(3-n)}\left[\frac{2}{3-n}\sqrt{\tilde{s}}\right] \bullet K_{\frac{1-n}{3-n}}\left[\frac{2}{3-n}\sqrt{\tilde{s}}r_{eD}^{(3-n)/2}\right] + K_{2/(3-n)}\left[\frac{2}{3-n}\sqrt{\tilde{s}}\right] \bullet I_{\frac{1-n}{3-n}}\left[\frac{2}{3-n}\sqrt{\tilde{s}}r_{eD}^{(3-n)/2}\right]\right\}\right)} \tag{38}$$

Using the solution provided by [8], [4] presented pressure and pressure derivative plots for such behaviors as shown in Figs. 9 and 10. In these plots it is seen for closed systems in both pseudoplastic and dilatant cases, that the late-time pressure derivative behavior always displays a unit-slope line as for Newtonian fluids. As for Newtonian behavior, the late-time pressure derivative decreases in both dilatant or pseudoplastic cases.

[4] rewrote Equation 6 based on reservoir drainage area, so that:

$$t_{DA} = \frac{t}{G\left(\pi r_e^{3-n}\right)} \tag{39}$$

[4] combined Equations 11, 31 and 39 to develop an analytical expression to find well drainage area,

$$A = \pi\left[\frac{t_{rpiNN}}{G} * \left(\frac{1}{4}\right)^{1/\alpha - 1}\right]^{2/3-n} \tag{40}$$

Where t_{rpiNN} is the intersection point formed by the straight-lines of the radial and pseudosteady-state flow regimes. The above equation was multiplied by $(\pi^{(1/\alpha-1)})^{1/3-n}$ as a correction factor. This is valid for both dilatant and pseudoplastic non-Newtonian fluids.

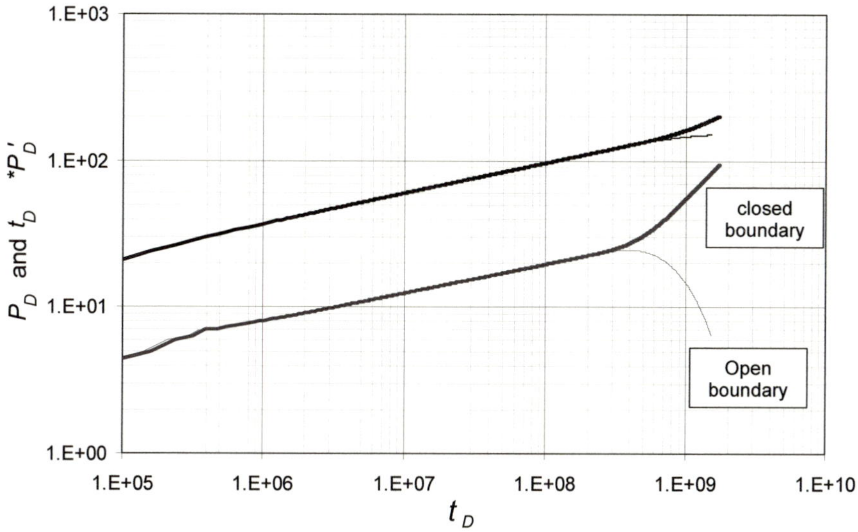

Figure 9. Dimensionless pressure and pressure derivative behavior in closed and open boundary systems for a non-Newtonian pseudoplastic fluid with $n = 0.5$, $r_e = 2000$ ft

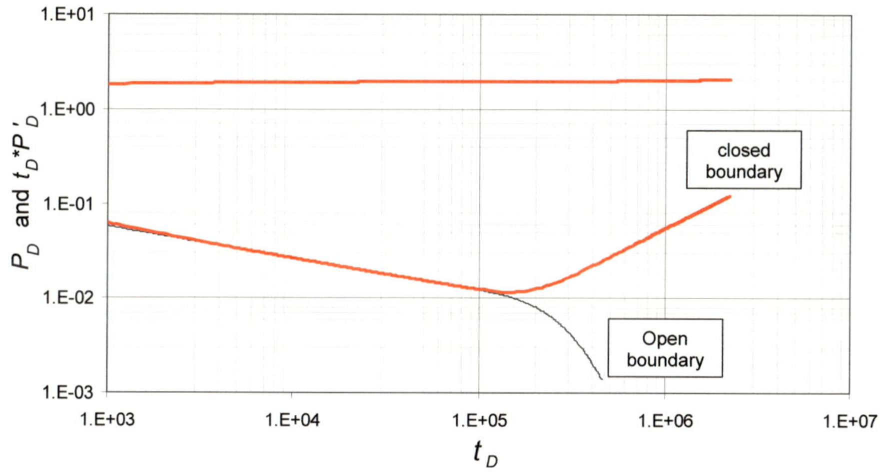

Figure 10. Dimensionless pressure and pressure derivative behavior in closed and open boundary systems for a non-Newtonian dilatant fluid with $n = 1.5$, $r_e = 2000$ ft

There is no pressure derivative expression for open boundary systems. Then, for pseudoplastic fluids the following correlation was also developed [4],

$$t_{DA_{NN}} = -0.003n^2 + 0.0337n + 0.052 \tag{41}$$

Equating Equation 41 to 39 and solving for reservoir drainage area, such as:

$$A=\pi\left[\frac{t_{rsiNN}}{G\pi\left(-0.003n^{2}+0.0337n+0.052\right)}\right]^{2/3-n} \tag{42}$$

For dilatant fluids the correlation found is:

$$t_{DA_{NN}}=0.9175n^{3}-3.7505n^{2}+5.1777n-2.2913 \tag{43}$$

In a similar fashion as for the pseudoplastic case,

$$A=\pi\left[\frac{t_{rsiNN}}{G\pi\left(0.9175n^{3}-3.7505n^{2}+5.1777n-2.2913\right)}\right]^{2/3-n} \tag{44}$$

t_{rsiNN} in Equations 42 and 44 corresponds is the intersection point formed by the straight-line of the radial and negative unit-slope line drawn tangentially to the steady-state flow regime.

Example 3. [4] presented a synthetic example to determine the well drainage area. Pressure and pressure derivative data are provided in Figure 11 and other relevant information is given below:

$n = 0.5$	$h = 16.4$ ft	$k = 350$ md	$q = 300$ BPD
$\phi = 5$ %	$B_o = 1$ rb/STB	$\mu_{eff} = 0.014833$ cp	$c_t = 0.0000689$ psi^{-1}
$r_w = 0.33$ ft	$H = 20$ cp*s^{n-1}	$r_e = 2000$ ft	$P_i = 2500$ psi

Solution. From Figure 11, the intercept point, t_{rpiNN}, of the radial and pseudosteady-state straight lines is 60 hr which is used in Equation 40 to provide a well drainage area of 275 acres. Notice that this reservoir has an external radius of 2000 ft which represents an area of 288 acres. This allows obtaining an absolute error of 2.33 %.

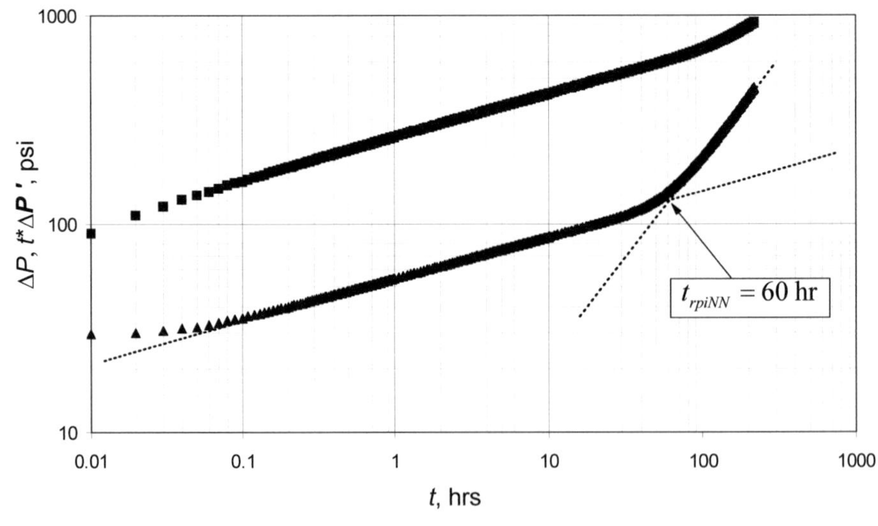

Figure 11. Pressure and pressure derivative for example 3

7. Heterogeneous reservoirs

In the well interpretation area of the Petroleum Engineering discipline a homogeneous reservoir is conceived to possess a single porous matrix while a heterogeneous reservoir has a porous matrix and either vugs or fractures. A common term used for heterogeneous systems is naturally-fractured reservoirs. However, this term is not recommended to be used since the fractures may result for either a mechanic process or a chemical process (matrix dissolution). Therefore, a more convenient term used in this book is double porosity systems in which the well is fed by the fractures and the fractures are fed by the matrix. By the same token, in a double-permeability system the well is fed by both fractures and matrix and the fractures are also fed by the matrix. This last one, however, has little application in the oil industry.

The governing well pressure solution in the Laplacian domain for a double-porosity system with a non-Newtonian fluid excluding wellbore storage and skin effects was provided by [19] as:

$$\tilde{P}_{DNN} = \frac{K_{\frac{1-n}{3-n}}\left(\frac{2}{3-n}\sqrt{\tilde{s}f(\tilde{s})}\right)}{\tilde{s}\left(\sqrt{\tilde{s}f(\tilde{s})}K_{\frac{2}{3-n}}\left(\frac{2}{3-n}\sqrt{\tilde{s}f(\tilde{s})}\right)\right)} \tag{45}$$

The Laplacian parameter, $f(\tilde{s})$ is a function of the model type and fracture system geometry and is given by:

$$f(s) = \frac{\omega(1-\omega)s+\lambda}{(1-\omega)s+\lambda} \tag{46}$$

[2] implemented the *TDS* methodology for characterization of double-porosity systems with pseudoplastic fluids. As for Newtonian case, the infinite-acting radial flow regime is represented by a horizontal straight line on the pressure derivative curve. The first segment corresponds to pressure depletion in the fracture network while the second portion is due to the pressure response of an equivalent homogeneous reservoir. On the other hand, the transition period which displays a trough on the pressure derivative curve during the transition period depends only on the dimensionless storage coefficient, ω. The warren and Root parameters are defined in reference [26].

Figure 12 shows a log-log plot of the dimensionless pressure and pressure derivative for a double- porosity system with constant interporosity flow parameter, constant n value and variable dimensionless storage coefficient the higher ω the less pronounced the trough. As seen there, as the value of n decreases, the slope of the derivative during radial flow increases. In Figure 13 is shown the effect of variable of the interporosity flow parameter for constant values of dimensionless storage coefficient and flow behavior index. Notice in that plot that as the value of λ decreases, the transition period shows up later. Finally, Figure 14 shows the effect of changing the value of the flow behavior index for constant values of λ and ω. The effect of the increasing the pressure derivative curve's slope is observed as the value of n decreases. Needless to say that neither wellbore storage nor skin effects are considered.

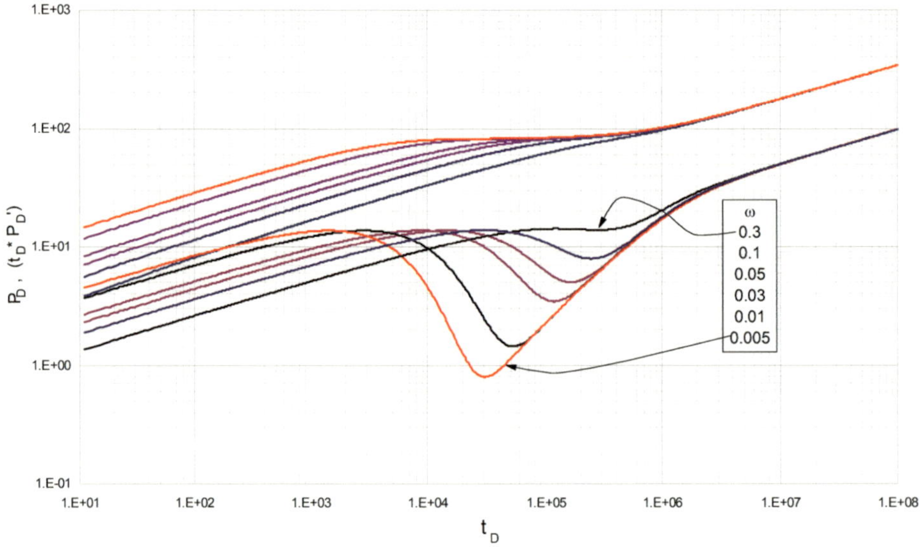

Figure 12. Dimensionless pressure and pressure derivative log-log plot for variable dimensionless storage coefficient, λ=1x10^{-6} and n=0.2 for a heterogeneous reservoir

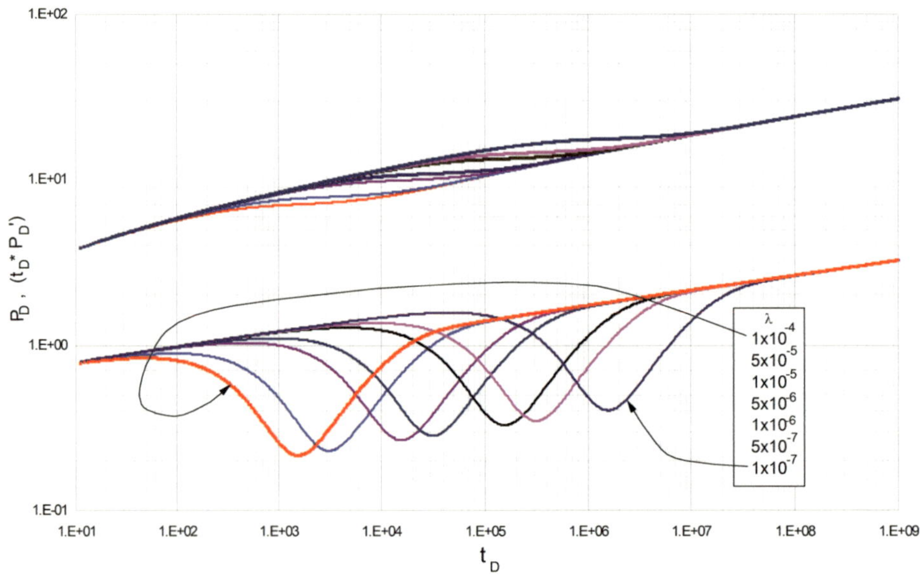

Figure 13. Dimensionless pressure and pressure derivative log-log plot for variable interporosity flow parameter, ω=0.05 and n=0.8 for a heterogeneous reservoir

The infinite-acting radial flow regime is identified by a straight line which slope increase as the value of the flow behavior index decreases. See Figure 14. The first segment of such line corresponds to the fracture-network dominated period, and, the second one -once the transition effects are no longer present-, responds for a equivalent homogeneous reservoir. An expression for the slope is given [11] as:

$$m = \frac{n-1}{n-3} \tag{47}$$

Also, the slope of the pressure derivative during radial flow regime is related to the flow behavior index by:

$$n = -1.8783425 - 7.8618321m^3 + 0.19406557m^{0.5} + 2.8783425e^{-m} \tag{48}$$

As observed in Figure 12, as the dimensionless storage coefficient decreases the transition period is more pronounced no matter the value of the interporosity flow parameter. Therefore, a correlation for $0 \le \omega \le 1$ with an error lower than 3 % as a function of the minimum time value of the pressure derivative during the trough, the flow behavior index and the beginning of the second of the infinite-acting radial flow regime is developed in this study as:

$$\frac{1}{\omega} = \left| 3180.6369 + 551.0582\left(\ln\frac{t_{min}}{t_{b2}}\right)^2 - \frac{2053.5888}{x^{0.5}} + \frac{75.337547}{x} - \frac{1.4787073}{x^{1.5}} - \frac{910.05377}{n^{0.5}} + \frac{988.80592}{n} - \frac{459.61296}{n^{1.5}} + \frac{73.93695}{n^2} \right| \tag{49}$$

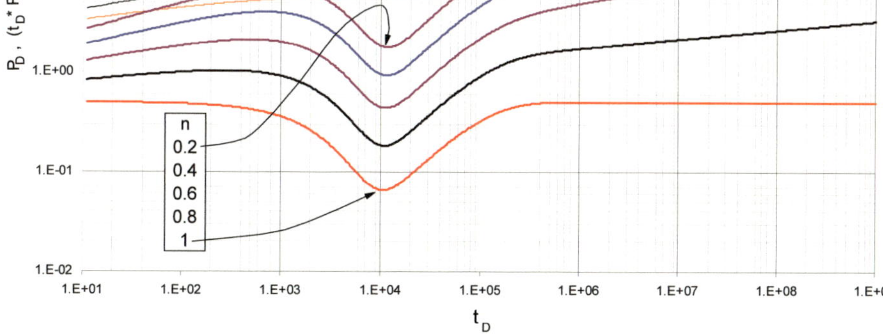

Figure 14. Dimensionless pressure and pressure derivative log-log plot for variable flow behavior index, ω=0.03 and λ=1x10^{-5} for a heterogeneous reservoir

Another way to estimate ω uses a correlation which is a function of the intersection time between the unit-slope pseudosteady-state straight line developed during the transition period, the time of the trough. We also found that this correlation is also valid for $0 \le \omega \le 1$ with an error lower than 0.7 %.

$$\omega = 0.019884508 - \frac{1.153351}{y} + \frac{43.428536}{y^2} - \frac{555.85387}{y^3} + \frac{3232.6805}{y^4} - \frac{6716.9801}{y^5} - \frac{0.0093613189}{n} + \frac{0.0042870178}{n^2} + \frac{0.00027356586}{n^3} - \frac{0.0005221335}{n^4} + \frac{0.000072466135}{n^5} \quad (50)$$

A final correlation to estimate ω valid for $0 \le \omega \le 1$ with an error lower than 0.4 % is given as follows:

$$\omega = \frac{-0.098427346 + 0.00046337048y + 0.000025063353y^2 - 0.00000050316996y^3 + 0.0036057682n - 0.0073959605n^2}{1 - 0.36468068y - 0.064934748n - 0.047596083n^2} \quad (51)$$

The interporosity flow parameter also plays an important role in the characterization of double porosity systems. From Figure 13, it is observed that the smaller the value of λ the later the transition period to be shown up. A correlation for it was obtained using the time at the trough and the dimensionless storage coefficient, as presented by next expression:

$$\lambda = \frac{\left(6.9690127\times10^{-7} + 3.4893658\times10^{-8}n - 3.2315082\times10^{-8}n^2 - 5.9013807w + 21571690w^2 + 3.6102987\times10^{12}w^3\right)}{\left(1 + 0.0099353372n - 3740035.1w + 6.7143604\times10^{12}w^2\right)} \quad (52)$$

Equation 51 is valid for $1\text{x}10^{-4} < \lambda < 9\text{x}10^{-7}$ with an error lower than 4 %. A correlation involving the coordinates of the trough is given as:

$$\begin{aligned}\lambda = &-0.00082917155 - 0.0014247498n - 0.00028717451 \\ &z - 0.00077173053n^2 - 3.2538271\times10^{-5}z^2 - 0.0003203949nz - \\ &-0.0001423889n^3 - 1.212213\times10^{-6}z^3 - 1.7831692\times10^{-5}nz^2 - \\ &-8.6457217\times10^{-5}n^2 z\end{aligned} \quad (53)$$

Which is valid for $1\text{x}10^{-4} < \lambda < 9\text{x}10^{-7}$ with an error lower than 3.7 %. Another expression for λ within the same mentioned range involving the minimum time of the trough is given for an error lower than 1.3 %.

$$\ln\lambda = -2.1223034 - 0.09473309n + 0.077489686n^{0.5}\ln(n) - \frac{0.010651118}{n^{0.5}} - \frac{0.043958503}{w^{0.5}} + \frac{1.5653137\times10^{-5}\ln w}{w} + \frac{0.00024143014}{w} + \frac{8.7148736\times10^{-9}}{w^{1.5}} - \frac{4.0331364\times10^{-13}}{w^2} \quad (54)$$

Example 4. Figure 15 contains the pressure and pressure derivative log-log plot of a pressure test simulated by [2] with the information given below. It is requested to estimate from these data the dimensionless storage coefficient and the interporosity flow parameter.

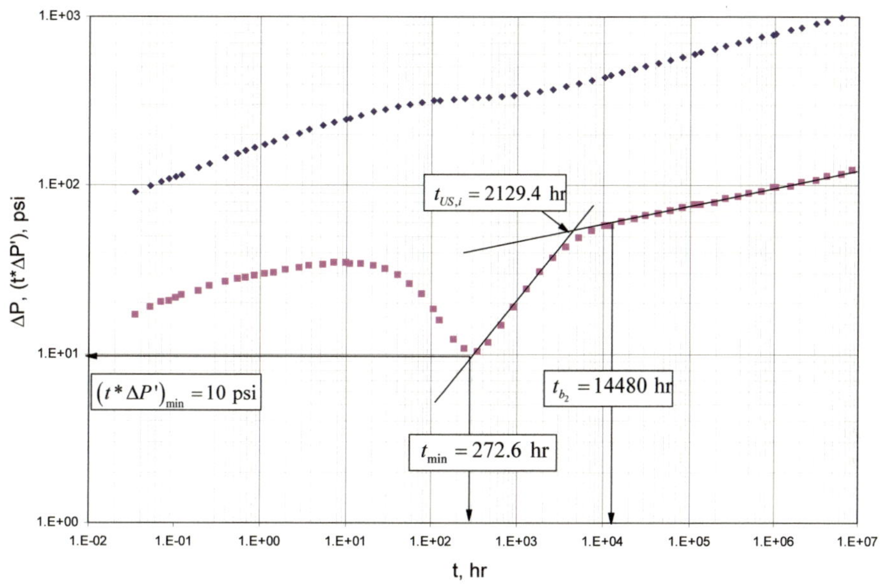

Figure 15. Pressure and pressure derivative for example 4

Solution. From Figure 15 the following characteristic points are read:

t_{min} = 272.6 hr t_{b2} = 14480 hr $t_{US,i}$ =2129.4 hr $(t^*\Delta P')_{min}$ = 10 psi

Using Equations 5 and 6, the above data are transformed into dimensionless quantities as follows:

t_{Dmin} =32000 t_{Db2} = 17000000 $t_{DUS,i}$ =250000 $(t_D{}^*P_D')_{min}$ =0.31

During the infinite-acting radial flow regime the following points were arbitrarily read:

$(t)_{r1}$ = 35724.9 hr $(t^*\Delta P')_{r1}$ = 67.292 psi $(t)_{r2}$ =56169.5 hr $(t_D{}^*\Delta P')_{r2}$ = 61.2283 psi

With these points a slope is estimated to be m = 0.108. Equation 47 allows obtaining a flow behavior index of 0.76. The Warren and Root's naturally fractured reservoir parameters are estimated as follows:

Equation	ω	Equation	λ
49	0.052	52	5.01E-06
50	0.05	53	5.043E-06
51	0.05	54	3.66E-06

Table 1. Summary of results for example 4

As a final remark, I would like to comment that some crude oils or other type of fluids used in the oil industry may display a non-Newtonian Bingham-type behavior. It is common to deal with Non Newtonian fluids during fracturing and drilling operations and oil recovery processes, as well. When a reservoir contains a non-Newtonian fluid, such as those injected during EOR with polymers flooding or the production of heavy-oil, the interpretation of a pressure test for these systems cannot be conducted using the conventional models for Newtonian fluid flow since it will lead to erroneous results due to a completely different behavior.

The problem considered now, presented in reference [27], involves the production of a Bingham fluid from a fully penetrating vertical well in a horizontal reservoir of constant thickness; the formation is saturated only with the Bingham fluid. The basic assumptions are: (a) Isothermal, isotropic and homogeneous formation, (b) Single-phase horizontal flow without gravity effects, (c) Darcy's law applies, and (d) Constant fluid properties and formation permeability.

The governing flow equation can be derived by combining the modified Darcy's law with the continuity equation and is expressed in a radial coordinate system as:

$$\frac{k}{r}\frac{\partial}{\partial r}\left[\frac{\rho(P)}{\mu_B}r\left(\frac{\partial P}{\partial r}-G\right)\right]=\frac{\partial}{\partial t}\left[\rho(P)\phi(P)\right] \tag{55}$$

The density of the Bingham fluid, $\rho(P)$, and the porosity of the formation, $\phi = \phi(P)$, are functions of pressure only, so Equation 54 may be rewritten as:

$$\frac{1}{r}\frac{\partial}{\partial r}\left[r\left(\frac{\partial P}{\partial r}-G\right)\right]=\frac{\phi\mu_B c_t}{k}\frac{\partial P}{\partial t} \tag{56}$$

The initial condition is:

$$P(r,t=0)=P_i, \qquad r\geq r_w \tag{57}$$

At the wellbore inner boundary, $r = r_w$, the fluid is produced at a given production rate, q; then, the inner boundary condition is:

$$q=2\pi rh\frac{k}{\mu_B}\left(\frac{\partial P}{\partial r}-G\right)_{r=r_w} \tag{58}$$

Parameter G is de minimum pressure gradient which expressed in dimensionless form yields:

$$G_D=\frac{Gr_w kh}{141.2q\mu_B B} \tag{59}$$

[15] solved numerically Equation 55 and provided an interpretation technique for this type of fluids using the pressure and pressure derivative log-log plot. For a Bingham-type non-Newtonian fluid, this behavior changes by observing that there is a point where the dimensionless pressure derivative is high and this increases with an increase of G_D and the reservoir radius, Figure 16.

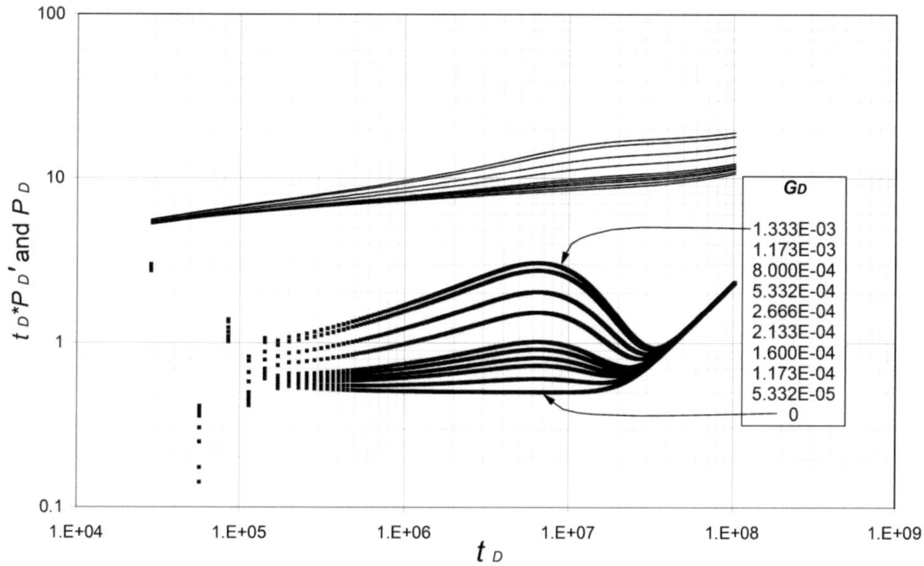

Figure 16. Dimensionless pressure and derivative pressure for $r_{eD} = 9375$

8. Conclusion

This chapter comprises the most updated state-of-the-art for well test interpretation in reservoirs having a non-Newtonian fluid. Extension of the *TDS* technique along with practical examples is given for demonstration purposes. This should be of extreme importance since most heavy oil fluids behave non-Newtonially, then, its characterization using conventional analysis is inappropriate and the methodology presented here are strongly recommended.

Nomenclature

B	Volumetric factor, RB/STB
c_t	System total compressibility, 1/psi
C	Wellbore storage, bbl/psi
C_{fD}	Dimensionless fracture conductivity

h	Formation thickness, ft
H	Consistency (Power-law parameter), cp*s^{n-1}
G	Group defined by Equation 3
G	Minimum pressure gradient, Psi/ft
G_D	Dimensionless pressure gradient
k	Permeability, md
k	Flow consistency parameter
$k_f w_f$	Fracture conductivity, md-ft
m	Slope
n	Flow behavior index (power-law parameter)
P	Pressure, psi
q	Flow/injection rate, STB/D
t	Time, hr
r	Radius, ft
$t*\Delta P'$	Pressure derivative, psi
t_D*P_D'	Dimensionless pressure derivative
r_a	Distance from well to non-Newtonian/Newtonian front/interface
w	ω/t_{Dmin}
$\tilde{s}$	Laplace parameter
x	t_{min}/t_{b2}
x_f	Half-fracture length, ft
y	t_{USi}/t_{min}
z	$\ln[(t_D*P_D')_{\min}/t_{Dmin}]$

Table 2. Nomenclature of main variables

Δ	Change, drop
ϕ	Porosity, Fraction
γ	Shear rate, s^{-1}
σ	Shear stress, N/m
λ	Dimensionless interposity parameter
μ	Viscosity, cp
μ_{eff}	Effective viscosity for power-law fluids, cp*$(s/ft)^{n-1}$
μ_B	Bingham plastic coefficient, cp
μ^*	Characteristic viscosity, cp/ft^{1-n}
τ	Shear stress, N/m
ω	Dimensionless storativiy coefficient

Table 3. Greeks

1	Non-Newtonian region
2	Newtonian region
app	Apparent
D	Dimensionless
DA	Dimensionless based on area Based on area
Dxf	Dimensionless based on half-fracture length Based on area
e	External
eff	Effective
i	Initial
LPi	Intersect of linear and pseudosteady-state lines
M	Maximum
N	Newtonian
NN	Non-Newtonian
r	Radial (any point on radial flow)
RLi	Intersect of radial and linear lines
rpiNN	Intersect of radial and pseudosteady-state lines
rsiNN	Intersect of radial and steady-state lines
w	Wellbore

Table 4. Suffices

Author details

Freddy Humberto Escobar
Universidad Surcolombiana, Neiva (Huila), Colombia, South America

9. References

[1] Escobar, F.H., Martinez, J.A., and Montealegre-M., Matilde, 2010. Pressure and Pressure Derivative Analysis for a Well in a Radial Composite Reservoir with a Non-Newtonian/Newtonian Interface. CT&F. Vol. 4, No. 1. p. 33-42. Dec. 2010.

[2] Escobar, F.H., Zambrano, A.P, Giraldo, D.V. and Cantillo, J.H., 2011. Pressure and Pressure Derivative Analysis for Non-Newtonian Pseudoplastic Fluids in Double-Porosity Formations. CT&F, Vol. 5, No. 3. p. 47-59. June.

[3] Escobar, F.H., Bonilla, D.F. and Ciceri, Y,Y. 2012. Pressure and Pressure Derivative Analysis for Pseudoplastic Fluids in Vertical Fractured wells. Paper sent to CT&F to request publication.

[4] Escobar, F.H., Vega, L.J. and Bonilla, L.F., 2012b. Determination of Well Drainage Area for Power-Law Fluids by Transient Pressure Analysis". Paper sent to CT&F to request publication.

[5] Fan, Y. 1998. A New Interpretation Model for Fracture-Calibration Treatments. SPE Journal. P. 108-114. June.

[6] Hirasaki, G.J. and Pope, G.A., 1974. Analysis of Factors Influencing Mobility and Adsorption in the Flow of Polymer Solutions through Porous Media. Soc. Pet. Eng. J. Aug. 1974. p. 337-346.

[7] Igbokoyi, A. and Tiab, D., 2007. New type curves for the analysis of pressure transient data dominated by skin and wellbore storage: Non-Newtonian fluid. Paper SPE 106997 presented at the SPE Production and Operations Symposium, 31 March – 3 April, 2007, Oklahoma City, Oklahoma.

[8] Ikoku, C.U. 1978. Transient Flow of Non-Newtonian Power-Law Fluids in Porous Media. Ph.D. dissertation. Stanford U., Stanford, CA.

[9] Ikoku, C.U., 1979. Practical Application of Non-Newtonian Transient Flow Analysis. Paper SPE 8351 presented at the SPE 64th Annual Technical Conference and Exhibition, Las Vegas, NV, Sept. 23-26.

[10] Ikoku, C.U. and Ramey, H.J. Jr., 1979. Transient Flow of Non-Newtonian Power-law fluids Through in Porous Media. Soc. Pet. Eng. Journal. p. 164-174. June.

[11] Ikoku, C.U. and Ramey, H.J. Jr., 1979. Wellbore Storage and Skin Effects During the Transient Flow of Non-Newtonian Power-law fluids Through in Porous Media. Soc. Pet. Eng. Journal. p. 164-174. June.

[12] Katime-Meindl, I. and Tiab, D., 2001. Analysis of Pressure Transient Test of Non-Newtonian Fluids in Infinite Reservoir and in the Presence of a Single Linear Boundary by the Direct Synthesis Technique. Paper SPE 71587 prepared for presentation at the 2001 SPE Annual Technical Conference and Exhibition held in New Orleans, Louisiana, 30 Sept.–3 Oct.

[13] Lund, O. and Ikoku, C.U., 1981. Pressure Transient Behavior of Non-Newtonian/Newtonian Fluid Composite Reservoirs. Society of Petroleum Engineers of AIME. p. 271-280. April.

[14] Mahani, H., Sorop, T.G., van den Hoek, P.J., Brooks, A.D., and Zwaan, M. 2011. Injection Fall-Off Analysis of Polymer Flooding EOR. Paper SPE 145125 presented at the

SPE Reservoir Characterization and Simulation Conference and Exhibition held in Abu Dhabi, UEA, October 9-11.

[15] Martinez, J.A., Escobar, F.H., and Montealegre-M, M. 2011. Vertical Well Pressure and Pressure Derivative Analysis for Bingham Fluids in a Homogeneous Reservoirs. Dyna, Year 78, Nro. 166, p.21-28. Dyna. 2011.

[16] Martinez, J.A., Escobar, F.H. and Cantillo, J.H., 2011b. Application of the *TDS* Technique to Dilatant Non-Newtonian/Newtonian Fluid Composite Reservoirs. Article sent to Ingeniería e Investigación to request publication. Ingeniería e Investigación. Vol. 31. No. 3. P. 130-134. Aug.

[17] Martinez, J.A., Escobar, F.H. and Bonilla, L.F. 2012. "Numerical Solution for a Radial Composite Reservoir Model with a No-Newtonian/Newtonian Interface. Paper sent to CT&f to request publication.

[18] Odeh, A.S. and Yang, H.T., 1979. Flow of non-Newtonian Power-Law Fluids Through in Porous Media. Soc. Pet. Eng. Journal. p. 155-163. June.

[19] Olarewaju, J.S., 1992. A Reservoir Model of Non-Newtonian Fluid Flow. SPE paper 25301.

[20] Savins, J.G., 1969. Non-Newtonian flow Through in Porous Media. Ind. Eng. Chem. 61, No 10, Oct. 1969. p. 18-47.

[21] Tiab, D., 1993. Analysis of Pressure and Pressure Derivative without Type-Curve Matching: 1- Skin and Wellbore Storage. Journal of Petroleum Science and Engineering, Vol 12, pp. 171-181.Also Paper SPE 25423, Production Operations Symposium held in Oklahoma City, OK. pp 203-216.

[22] Tiab, D., Azzougen, A., Escobar, F. H., and Berumen, S. 1999. *"Analysis of Pressure Derivative Data of a Finite-Conductivity Fractures by the 'Direct Synthesis Technique'."* Paper SPE 52201 presented at the 1999 SPE Mid-Continent Operations Symposium held in Oklahoma City, OK, March 28-31, 1999 and presented at the 1999 SPE Latin American and Caribbean Petroleum Engineering Conference held held in Caracas, Venezuela, 21–23 April.

[23] van Poollen, H.K., and Jargon, J.R. 1969. Steady-State and Unsteady-State Flow of Non-Newtonian Fluids Through Porous Media. Soc. Pet. Eng. J. March 1969. p. 80-88; Trans. AIME, 246.

[24] Vongvuthipornchai, S. and Raghavan, R. 1987. Pressure Falloff Behavior in Vertically Fractured Wells: Non-Newtonian Power-Law Fluids. SPE Formation Evaluation, December, p. 573-589.

[25] Vongvuthipornchai, S. and Raghavan, R. 1987b. Well Test Analysis of Data Dominated by Storage and Skin: Non-Newtonian Power-Law Fluids. SPE Formation Evaluation, December, p. 618-628.

[26] Warren, J.E. and Root, P.J. 1963. The Behavior of Naturally Fractured Reservoirs. Soc. Pet. Eng. J. (Sept. 1963): 245-255.

[27] WU, Y.S., 1990. Theoretical Studies of Non-Newtonian and Newtonian Fluid Flow Through Porous Media. Ph.D. dissertation, U. of California, Berkeley.

Chapter 8

The Role of Geoengineering in Field Development

Patrick W. M. Corbett

Additional information is available at the end of the chapter

http://dx.doi.org/10.5772/50535

1. Introduction

The oil and gas industry is truly multi-disciplinary when it comes to analysing, modelling and predicting likely movement of fluids in the subsurface reservoir environment. Much has been written on the subject of integration in recent years and in this Chapter we can consider one particular approach to tackling the problem. The Petroleum Geoengineering[1] solution is offered to the origin, understanding, and static geological modelling of a reservoir and the simulation of the flow and the dynamic response to a production test. In field development these models remain a key monitoring and planning tool but here we consider the initial modelling steps only. As an example of building a heterogeneous reservoir model, we have chosen to illustrate this approach for certain types of fluvial reservoirs [which have presented challenges for reservoir description for many years, 2,3] which can benefit further from this detailed integrated approach. Furthermore, as such reservoirs are characterised by relatively low oil recovery, and where further intensive work by the industry will be needed to maintain hydrocarbon supplies in the future.

Integration challenge. It is often quoted that the use of the word "Integration" in SPE paper titles has followed a 'hockey stick' rise in recent years. Books have been written on the subject of integration and in the forward to one such study – Luca Cosentino[4] pointed out that studies were merely becoming less disintegrated as the industry evolved. The industry has developed ever more powerful, cross-disciplinary software platforms and workflows to help integration. In parallel is the need for professionals to stay abreast of the key work processes in each discipline and this chapter helps illustrate one such integrated approach from a scientific/technological approach rather than embedded in or wedded to particular software.

Geoengineering concept. This concept was introduced [1] into petroleum industry to capture the spirit of the workflow being a seamless progression from geological conceptual understanding, through petrophysical description to a numerical model and prediction of a dynamic response. The Petroleum Geoengineering approach outlined here is a small component of an all encompassing "Intentional manipulation of the subsurface environment as practiced by the petroleum industry with global impact". The recovery of oil and gas and the management of CO2 being the ultimate outcome and target of this approach.

Static and dynamic reservoir characterisation. Reservoir Characterisation is defined as the numerical quantification of reservoirs for numerical simulation. The petroleum industry often refers to static and dynamic characterisation of the subsurface and many workers will have their own interpretation of the terms. In the context of this Chapter we describe the rocks statically when we keep to a numerical characterisation of the rock at initial boundary conditions and dynamic being the response to some perturbation of the system (with production as an example). There are other definitions of static and dynamic properties (properties that can be changed versus those that cannot) but the above are followed here. Permeability – which only occurs during an experiment in response to a perturbation is considered static when it is the initial permeability of the system prior to the experiment.

Field Development: Field Development plans are based on computer simulation models of the field. This models consisting of multi-million cells are built by geologists for simulation by engineers. The resolution of geological models is often higher than can be accommodated by the flow simulation (particularly when complex fluids are involved). There is usually a reduction of geological detailed as the cells are upscaled in order to reduce the number of cells for computational expediency. The fundamental challenge being considered here is how detailed should the original model be and with this upscaling how the key properties are preserved in the model. Models are built prior to reservoir development, updated during the development and on continued use through the planning any improved oil recovery strategies and remain the key field development tool up until field abandonment.

2. Origin of reservoir heterogeneity

Clastic reservoirs are those made up of particles of rock that are the accumulated products of erosion, transport and deposition (Fig.1). Broadly speaking these are usually sands and clays and these types of reservoirs contain a significant proportion of the world's reservoirs. Within the clastic reservoir family are those reservoirs resulting from deposition by rivers – fluvial reservoirs. Fluvial reservoir types are very varied with braided and meandering being important end members. Depending on the slope, sediment supply, nature of the floodplain, rain fall and proximity to mountainous sediment sources, the resulting reservoir architecture will vary from low net:gross, meandering up to high net:gross, braided (Fig.2). This describes the macroscopic variation – but within the channel sand bodies are additional textural variations at various meso- and micro-scopic scales. Each of these scales will have a potential impact on the hydrocarbon recovery.

Role of texture in controlling reservoir properties. In sandstones, primary texture exerts a large influence of reservoir properties [5]. Primary texture is measured by grain size, grain shape, grain sorting, clay content, etc. Well sorted and rounded sands, in sandstone reservoirs, tend to have good porosity and high permeability. Fine grained sands of the same uniform shape and sorting will also have good porosity, but lower permeability. Poorly sorted sandstone with a variation in grain shape and size, will tend to have low porosity and permeability.

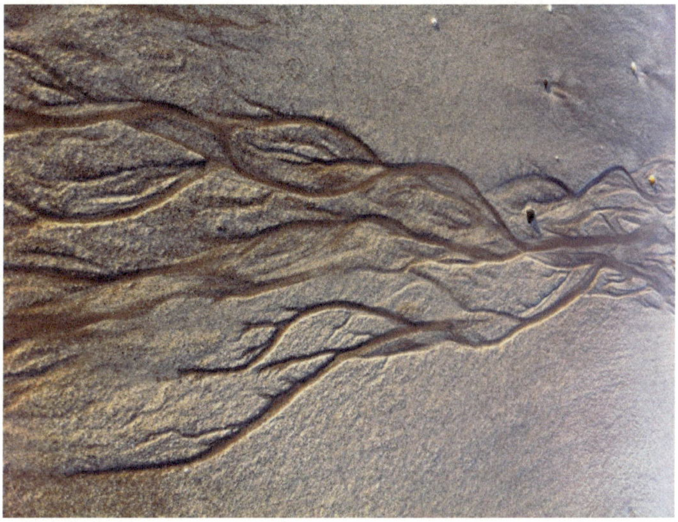

Figure 1. Modern analogue for a fluvial system showing characteristic channel channel complexity, Longcraigs Beach, Scotland

Figure 2. Various fluvial reservoirs at outcrop. Left low net:gross channelized, meandering, system from Tertiary, near Huesca, Spain. Right: High net:gross braided system from the Devonian, Scotland.

Where petrophysical heterogeneity in sandstones is present it is often due to the spatial distribution of these lithologies and their related properties which is why outcrop analogue studies remain a useful tool to define geobody geometries for reservoir modelling.

Use of Outcrop Analogues: The industry uses analogue reservoirs, outcropping on the surface, where relevant geological objects (geobodies) can be measured and their aspect ratios and stacking patterns determined (Fig. 3). Some very good outcrop analogues of fluvial reservoirs have been studies by the industry over the last 20 years [examples can be found over this period in 6,7] with outcrops in the UK (Yorkshire, Devon), Spain (S. Pyrenees) Portugal and the US (Utah) being used for reservoir studies in the North Sea, North Africa and Alaska. Geological age is not the critical consideration when it comes to chosing an analogue but net;gross (sand proportion in the system), channel size, bed load, flood plain, stacking patterns, climate, etc are more important criteria in selecting a 'good' outcrop analogue.

Figure 3. Outcrop of a fluvial system in Spain (near Huesca) where the average thickness of the channels was measured as 5.3m, with an average aspect ratio of 27:1 (with acknowledgement to the group of students who collected the data). Whilst only medium net:gross (35-45%) the channels are laterally stacked and within these layers the connectivity will be greater than expected from a simple model with a random distribution of sandbodies).

3. Measures of reservoir heterogeneity

Heterogeneity usually refers to variation of a property above a certain threshold (so as to distinguish from homogeneity). In reservoirs, the property we usually consider, when referring to heterogeneity, is that which controls flow, namely permeability. Porosity, which controls the hydrocarbon in place, in the fluvial reservoirs we are considering in this Chapter, tends by contrast to be relatively homogeneous. Heterogeneity can be described by statistical criteria from a sample data set. The petroleum geoengineer's starting point is often an analysis of heterogeneity. This determines what level of detail might be required to characterise the flow process. Heterogeneity is sometimes responsible for anisotropy – but not always – so we have also to consider this aspect of the reservoir's characteristics.

Porosity and permeability distributions. Heterogeneity is often first seen in a review of the histograms of the porosity and permeability data and these should always be part of an initial reservoir analysis.

Porosity data tends to form a symmetrical or normal distribution. Permeability on the other hand is often positively skewed, bimodal and usually highly variable (Fig. 4) [8]. It is a mistake to think of permeability as being always log-normally distributed (as is often implied in the literature) and the type of distribution should always be checked. Sometimes the distributions are clearly bi- (tri- or even multi-) modal and this aspect will require further analysis. Ideally each important element of the reservoir should be described by characteristic porosity and permeability distributions – and these can be used in the geological (i.e., geostatistical) modelling. Geostatistical (i.e., pixel) modelling is often performed in a Gaussian domain (Sequential Gaussian Simulation) and the skewed distributions are first transformed to Gaussian to make this technique most effective.

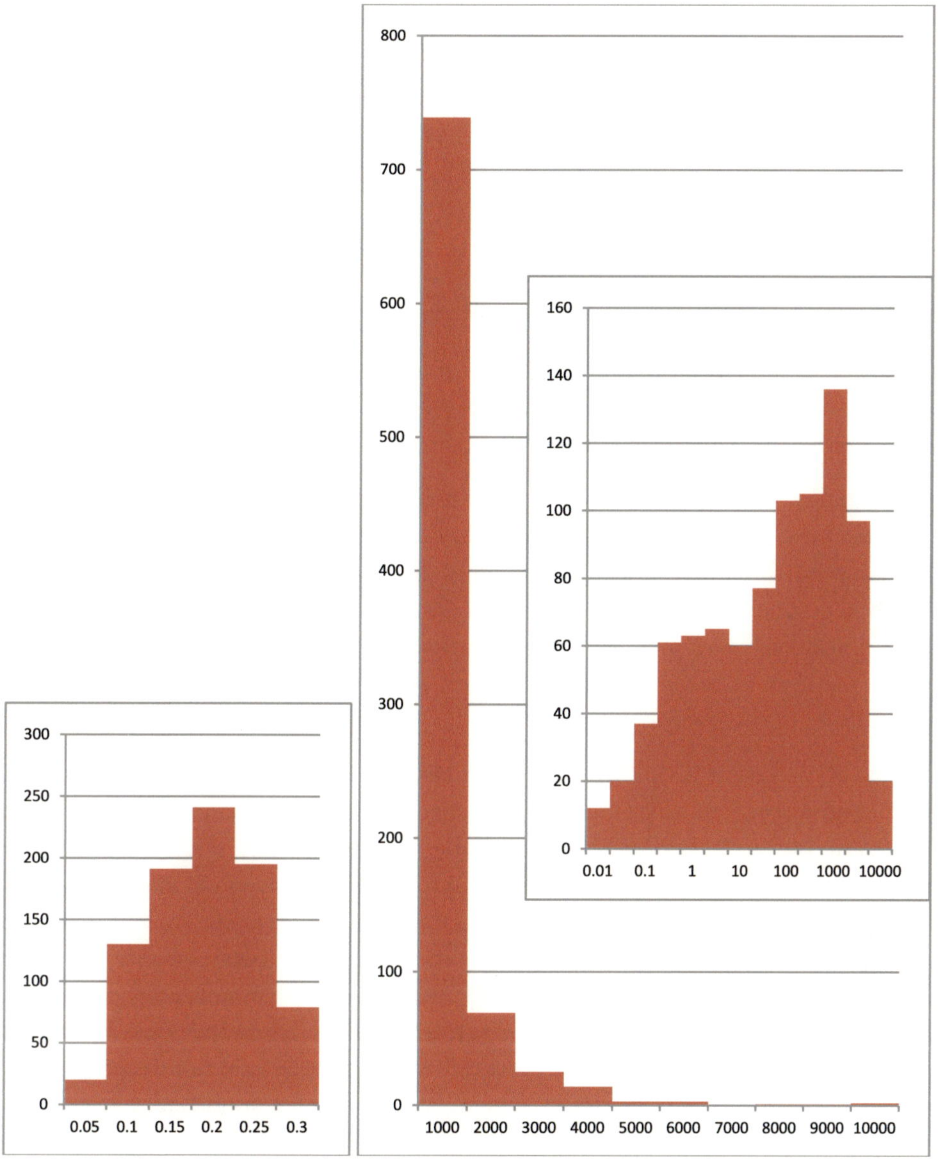

Figure 4. Porosity (Left – decimal), Permeability (Centre-mD) and Log Permeability (Right - mD) distributions for a fluvial data set. Porosity tends to a normal distribution with permeability being bimodal in the log domain (often this relates to properties of flood plain and channels)

Variation between Averages. Another useful indication of heterogeneity is apparent in differences between the arithmetic, geometric and harmonic averages (Table 1). For porosity these are often quite similar – but for permeability these can differ in fluvial reservoirs by orders of magnitude! Different averages have different applications in reservoir engineering and often used as a way of upscaling the directional flow properties (in the static model) in different directions.

	UK		North Africa	
Average	Poro	Perm	Poro	Perm
Arithmetic	0.167	441	0.108	25.7
Geometric	0.154	23.7	0.094	2.78
Harmonic	0.138	0.263	0.072	0.009

Table 1. Porosity (decimal) and permeability (mD) averages in fluvial sandstones – Left Triassic Sherwood Sandstone, UK (Fig.4); Right Triassic Nubian Sandstone, North Africa. Note relatively small differences between average porosity contrasting with order of magnitude variation between average permeabilities. This is further evidence of extreme permeability heterogeneity in these sandstones.

The arithmetic average is used as an estimator of horizontal permeability, and the harmonic for the vertical permeability, in horizontally layered systems. Where layered systems have different orientations (i.e., significant dip) then the averages need to be 'rotated' accordingly.

In the case of a random system, then the geometrical average is used in both horizontal and vertical directions. A truly random system, without any dominant directional structure, can also be assumed to be isotropic. Use of theses averages for upscaling comes with some caveats – the assumption that each data point carries the same weight (i.e., from a layer of the same thickness) and only single phase flow is being considered. In many fluvial reservoirs, the system is neither nicely layered nor truly random which requires careful treatment/use of the averages.

Coefficient of Variation (C_v). There are a number of statistical measures which are used in reservoir engineering to quantify the heterogeneity. The variance and the standard deviation are the well known ones used by all statisticians. However, in reservoir characterisation we tend to use the normalised standard deviation (standard deviation divided by the arithmetic average) as one such measure of heterogeneity and this is known as the Coefficient of Variation (Table 2) [8]. Another measure of heterogeneity, that probably has limited use to petroleum engineering only, is the Dykstra-Parsons coefficient (V_{DP}), but this assumes a log-normal distribution (of permeability) so tends to be used in modelling studies when a log-normal distribution is required to be input to the simulation process. The log-normal distribution, as discussed above, is not always found to be the case for permeability in reservoir rocks and therefore care has to be taken when using V_{DP}.

	UK		North Africa	
	Poro	Perm	Poro	Perm
S.D.	0.061	972	0.046	58.9
C_v	0.37	2.20	0.425	2.29

Table 2. Heterogeneity in porosity (decimal) and permeability (mD) averages in fluvial sandstones – Left Triassic Sherwood Sandstone, UK (Fig. 4); Right Triassic Nubian Sandstone, North Africa. Note porosity heterogeneity is low (but relatively high for sandstones) whereas permeability is very heterogeneous supporting the trend seen in the averages. Note these two Triassic reservoirs on different continents have remarkably consistent poroperm variability.

Lorenz Plot (LP): The Lorenz Plot (which is more widely known in economics as the GINI plot) is a specialised reservoir characterisation plot that shows the relative distributions of porosity and permeability in an ordered sequence (of high-to-low rock quality, essentially determined by the permeability, Fig. 5 right) and can be quantified – through the Lorenz Coefficient. Studying how porosity and permeability *jointly* vary is important. In Fig. 5 (left) 80% of the flow capacity (transmissivity) comes from just 30% of the storage capacity (storativity).

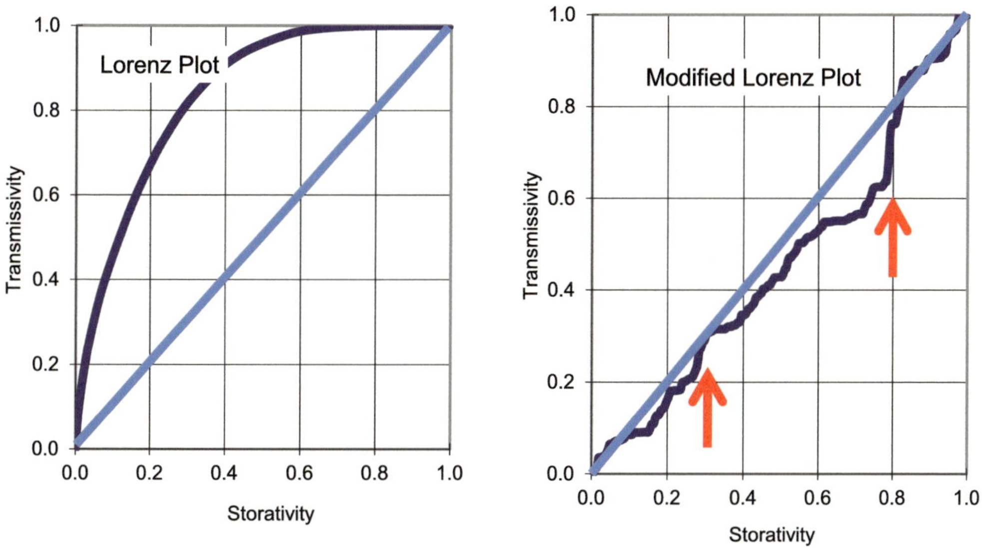

Figure 5. Example Lorenz and Modified Lorenz Plots for a fluvial data set (Fig.4). The LP (Right) shows high heterogeneity as the departure of the curve from the 45º line. The MLP (Left) shows presence of speed zones at various point (arrowed) in the reservoir. If the MLP is close to the 45º line then that is perhaps an indication of randomness and this can also be checked by variography.

The industry often uses cross plots of porosity and permeability – which will be discussed further below - which can focus the viewer on average porosity permeability relationships – but the LP should appear in every reservoir characterisation study as it emphasises the extremes that so often identify potential flow problems.

Modified Lorenz Plot (MLP). In a useful modification of the original Lorenz Plot where the re-ordering of the cumulative plot by original location provides the locations of the extremes (baffles and thief or speed zones). This plot (Fig. 5 left) has a similar profile to the production log and hence is an excellent tool for predicting inflow performance. The LP and MLP used in tandem can provide useful insights in to the longer term reservoir sweep efficiency and oil recovery.

Anisotropy vs Heterogeneity. With heterogeneity, sometimes comes anisotropy, particularly if the heterogeneity shows significant correlation structure. Correlation is measured by variography and where correlation lengths are different in different directions – this can identify anisotropy. Correlation in sedimentary rocks is often much longer in the horizontal and this gives rise to typical kv/kh anisotropy. In fluvial reservoirs, with common cross-bedding, the anisotropy often relates to small scale structure caused by the lamination but it is the larger scale connectivity that dominates (see the Exercise 1 in reference [1] for further consideration of this issue).

Rarely does significant anisotropy result from grain anisotropy alone as has been suggested by some authors. Anisotropy is a scale dependent property – smaller volumes tend to be isotropic (and this tendency is seen in core plugs) whereas at the formation scale bedding fabric tends to give more difference between kv and kh and therefore greater anisotropy. In fluvial systems, the arrangement of channel and inter-channel elements can have a significant effect on anisotropy. In high net:gross fluvial systems, well-intercalated channel systems will have higher tendency to be isotropic (geometric average) whilst preservation of more discrete channels will exhibit more anisotropic behaviour (arithmetic and harmonic average permeability). In this Chapter we are not considering natural or induced fractures which can increase anisotropy.

4. Reservoir rock typing

Petrophysicists use the term "Rock Typing" in a very specific sense - to describe rock elements (core plugs) with consistent porosity – permeability (i.e., constant pore size – pore throat) relationships. These relationships are demonstrated by clear lines on a poro-perm cross-plot and similar capillary pressure height functions. There are various ways these relationships can be captured (and the literature includes references to RQI, FZI, Amaefule, Pore radius, Winland, Lucia, RRT, GHE, Shenawi….) and each method directs the petrophysicist towards a consistent petrophysical sub-division of the reservoir interval. In Fig. 6 the coloured bands follow a consistent GHE approach based on the Amaefule FZI, RQI equation [9]. It matters not so much which rock typing method is used but that a rock typing method is used but that a rock typing method is used as the basis for reservoir description. Geologists and petrophyicists need to make these links work for an effective reservoir evaluation project. Special core analysis data when collected in a rock typing framework is most useful.

Property variation in poro-perm space. In fluvial reservoirs, it is very common to have a wide diversion of porosity and permeability (Fig. 6) due to the poorly sorted, immature, nature of these sands. Well sorted sands will have higher porosity and permeability than their poorly sorted neighbours. Coarse sands tend to have less primary clay content. Presence of mica and feldspar can also effect the textural properties – especially if the feldspar breaks down into clay components. Clays are more common in fine and poorly sorted sandstones. The variation of properties within fluvial systems is often a result of primary depositional texture. Diagenetic effects – especially where associated with calcrete (carbonate cement that is formed by surface evaporation and plant root influence in arid fluvial environments) or reworked calcrete into channel base (lag) deposits - can modify the original depositionally-derived properties (such as well cemented lag intervals) but perhaps do not change the overall permeability patterns. For this reason channel elements are often detected in fluvial reservoirs and are measured at outcrop for use in fluvial reservoir modelling studies.

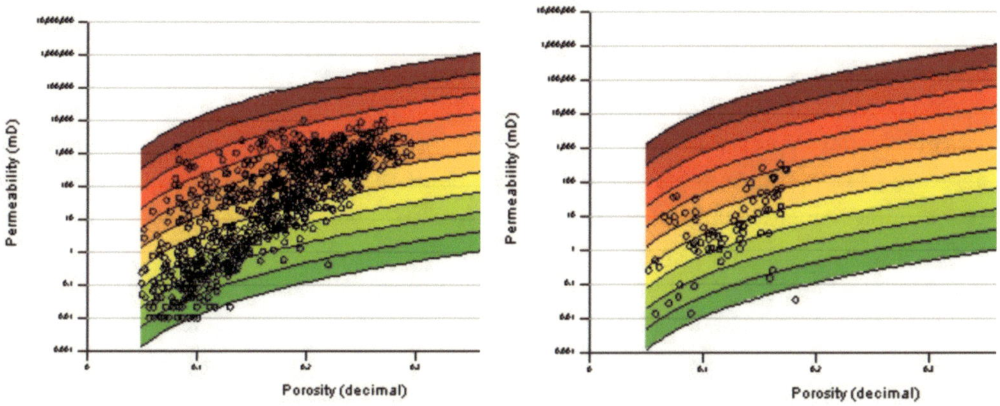

Figure 6. Porosity and permeability heterogeneity in fluvial sandstones – Left Triassic Sherwood Sandstone, UK; Right Triassic Nubian Sandstone, North Africa.

Link between geology and engineering. Rock types are a key link between geology and engineering as they are the geoengineering link between the depositional texture, the oil in place and the ease with which water can imbibe and displace oil. If fluvial reservoirs the presence of many rock types is critical to understanding oil-in-place and the, relatively low, recovery factors. Rock types are the fundamental unit of petrophysical measurement in a reservoir and failure to recognise the range of properties in a systematic framework can potentially result in the use of inappropriate average properties.

Link between MLP and rock typing. The MLP if coded by rock type can also emphasise the role of some rock types as conduits to flow and potential barriers/baffles to flow [10]. The link between rock types and heterogeneity is also important in understanding the "plumbing" in the reservoir – where are the drains, the speed zones, the thief zones, the baffles and the storage tanks?

Production logging. Ultimately the proof of what flows and what doesn't flow in a reservoir comes with the production (i.e, spinner) log. The spinner tool identifies flowing and non-flowing intervals (by the varying speed of rotation of a impellor in the well stream) and when correlated with the MLP can provide validation that the static and dynamic model are consistent [11]. If the best, and only the best, rock types are seen to be flowing then there is evidence of a double matrix porosity reservoir. If there is no correlation, then perhaps this points to evidence of a fractured (double porosity) system. The well test interpretation cannot distinguish between the two double porosity cases – but the production log perhaps can. Of course when it comes to interpreting downwhole data – there are also the downhole environment considerations needed (such as perforation location, perforation efficiency, water or gas influx, etc) to be taken into account. The geoengineering approach to calibrating a static model with a dynamic model for key wells (where there is perhaps core, log, production log and test data) and iterating until there's a match will have benefits when it comes to subsequent history matching of field performance.

Core to Vertical Interference Test comparison (k_v/k_h). Where there is also vertical interference data available, which is generally quite rarely, this can also be used to calibrate models of anisotropy [12]. The k_v/k_h ratio is often one of the critical reservoir performance parameters but rarely is there a comprehensive set of measurements. Core plug scale k_v/k_h measurements are not always helpful – as they are often 'contaminated' by local heterogeneity issues at that scale. Vertical plugs are often sampled at different – always wider – spacings, compared with horizontal plugs, and this means critical elements (which tend to be thin) controlling the effective vertical permeability are often missed. In fluvial reservoirs, these are often the overbank or abandonment shale intervals. Vertical plug measurements in shales are often avoided for pragmatic reasons (because measuring low permeability takes time and often the material doesn't lend itself to easy plugging). The effective k_v/k_h parameter that is needed for reservoir performance prediction often needs to be an upscaled measurement. Choosing the interval over which to conduct a representative vertical interference test is an important consideration if that route is chosen.

5. Dynamic well testing

Well testing is achieved by perforating, producing and shutting-in the well for a relatively short period of time, whilst recording the flow rates and (bottom-hole as the estimate of reservoir) pressures. The practical aspects are covered elsewhere in this book, here we consider the role of well test data in understanding the performance of fluvial reservoirs. The way that fluid flows towards the well bore following a perforating job, and the paths that the pressure drop takes in the reservoir are important considerations. Fluvial reservoirs are not homogeneous, isotropic, sands of constant thickness. They are systems with highly variable (showing many orders of magnitude permeability variation for the same porosity) internal properties. The paths (comprising both horizontal and vertical components) of pressure disturbance away from the well will depend very much on the 3D arrangement of the sand bodies and the floodplain characteristics – the reservoir plumbing (Fig.7) [13]. In this respect, fluvial reservoirs are some of the more complex (clastic) reservoirs encountered.

The diffusion of the pressure response into the reservoir is constrained by the diffusivity constant. In heterogeneous formations such as fluvial reservoirs this assumed constant isn't actually constant and varies with rock quality through the tested volume. In an ideal case the arithmetic average would be expected in the initial period of the test and the geometric at later stages (for a completely random system (Fig. 8). In reality, there are a number of less than ideal situations in the geology. Channels are not always big enough to see the first stabilisation clearly and the system is not absolutely random and therefore the geometric average is not always reached in the length of the test. These problems give rise to many well test interpretation challenges in fluvial reservoirs.

Cross-flow and comingled flow. When a reservoir is said to have cross-flow this means that the fluid passes in response to pressure changes between layers of different properties in the reservoir. This effect occurs in all directions – vertically and laterally – rather than in simple uniform radial directions from the well.

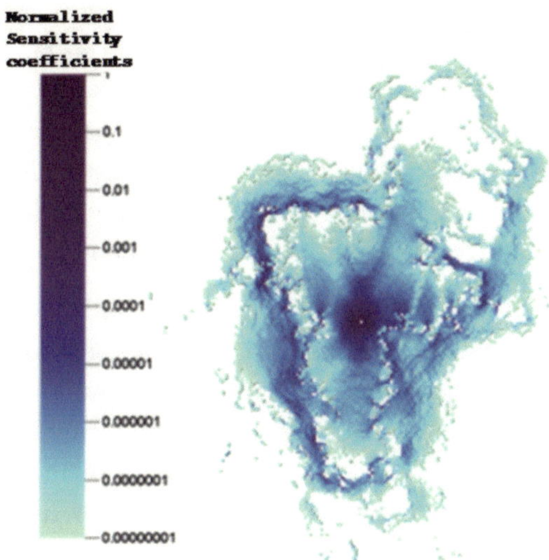

Figure 7. A simulation showing the location of the most sensitive parts of the formation at a particular time to the pressure response measured at the well. This effectively illustrates complex pressure diffusion (rather than simple radial flow) in a fluvial reservoir [13].

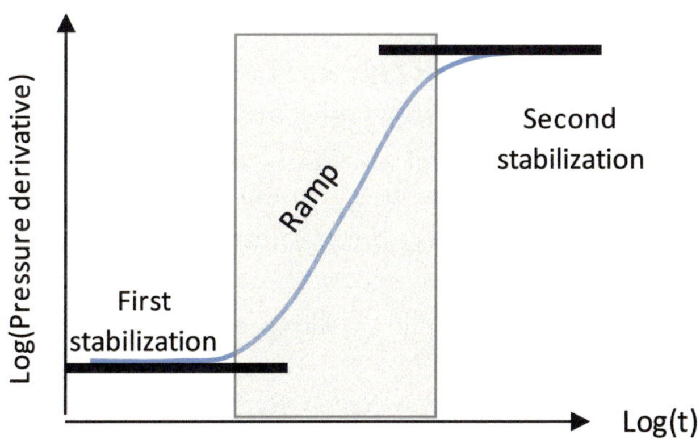

Figure 8. An ideal pressure derivative showing two stabilisations – the first would be expected to give the arithmetic average and the second, the geometric average. Remember that the difference between the arithmetic and geometric average in fluvial reservoirs is an order of magnitude or more (Table 1).

In a commingled reservoir the reservoir layers only communicate through the well bore. In the reservoir there is not flow between the layers. This situation is much more common in more layered reservoirs with laterally extensive shales between sheet-like (e.g. turbidite) sand bodies. Such situations can occur in fluvial systems – ephemeral channel sands sandwiching sheetflood deposits and interbedded shales – but perhaps as an exception, rather than the rule.

High net:gross fluvial reservoirs are often cross-flow in their internal drainage nature and cross-flow reservoirs are recognised as the most challenging for enhanced oil recovery. Gravity means that water slumps – or gas overrides – more easily in cross flow reservoirs. Shutting-off water influx – or gas – in producing well is ineffective as there are no laterally-extensive reservoir barriers present to base this strategy upon.

In homogeneous formations. Where the heterogeneity is low (C_v less than 0.5), the effects of cross flow are mitigated. Low heterogeneity fluvial sands can occur where the sands are relatively mature and far from source. This tends to occur in more distal locations. In these locations wind-blown sands can also occur and these are usually more uniform. In these situations well test will see the arithmetic (equals geometric) average permeability.

In heterogeneous formations. Where the heterogeneity is moderate (C_v between 0.5 and 1.0) these reservoirs might be dominated by cross bedding (not identified in low more homogeneous reservoirs) and these will induce strong capillary trapping. The well test might show reduced geometric average permeability in this case. Square root of k_x and k_y product for significant lateral (point bar) or downstream accretion-derived anisotropy.

In highly heterogeneous formations. Where the heterogeneity is very high (C_v greater than 1.0) and often this is the case with braided fluvial reservoirs then the most extreme cross flow can be seen. These are often detected by speed zones, drains) in the production log profile. Cross-flow introduces flow regime which can be confused with parallel (i.e. channel) boundaries. The ramp is seen best when the vertical permeability is effectively zero and the second stabilisation converges at the harmonic average **within** the commingled layers (Fig. 9 lower). The geometric average is seen when there is good connectivity and any channels near the well give rise to a geoskin response (Fig. 9 – top). In the middle case the restriction cause by the limits of the channels near the well is overcome in the later time by increased connectivity and this is the geochoke response (Fig. 9 - middle). These responses can be confused with the effects of faults (which may also be present and add to the confusion!).

Reservoir boundaries. The detection of reservoir boundaries is an important aspect of the well test interpretation. In relatively uniform sand properties then boundaries might be readily detected. In highly heterogeneous reservoirs cross flow effects might be misinterpreted as faults. It is often commented that well tests in fluvial reservoirs tend to show faults short (ca. 40ft) from the well. These may be channel margins or perhaps more likely, subtle, cross flow effects. The degree of heterogeneity is an important consideration in deducing boundaries (either sedimentological or structural) from internal cross flow effects. The impact of the two interpretations on the approach taken to reservoir modelling will be significant.

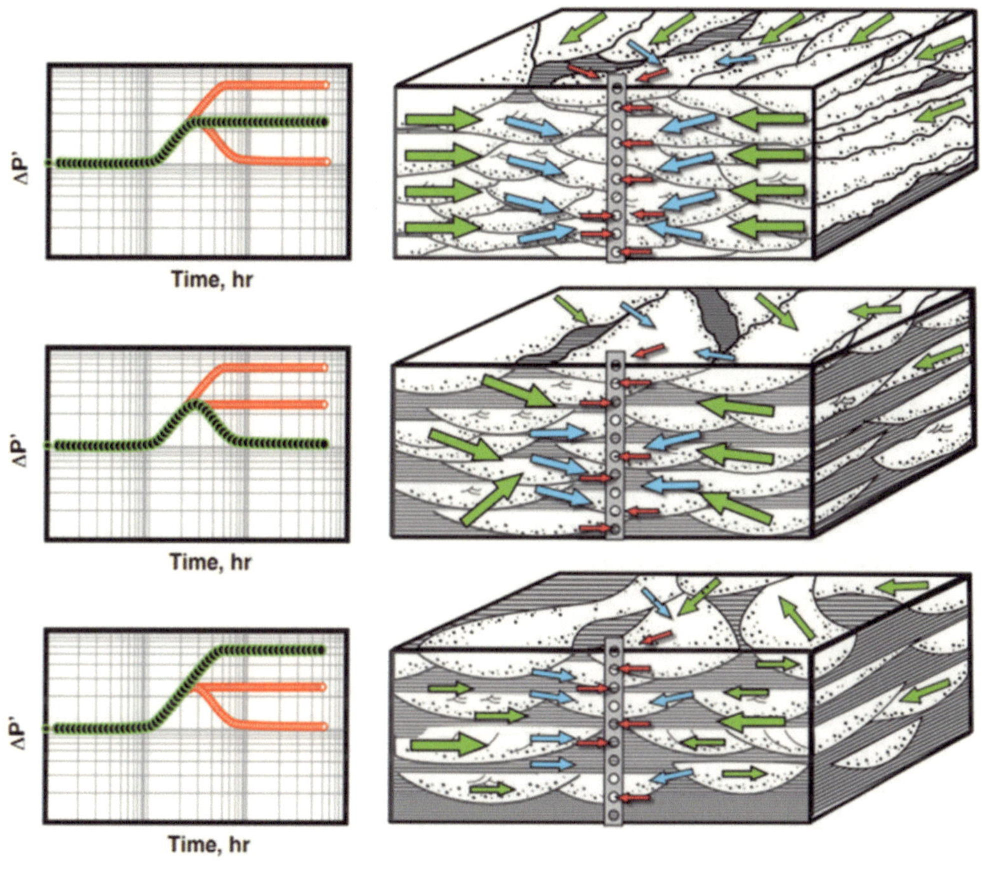

Figure 9. Shows various connectivity arrangements in fluvial reservoirs (between channels and floodplain) and an equivalent schematic pressure deriviative responses to the scenarios [13]. With subtle changes in lateral and vertical connectivity the response changes from a geoskin response (top) to a geochoke response (middle) or to a ramp response (lower).

Reservoir limit tests. Fluvial systems to produced sand bodies that are limited in extent (point bars). These are characterised but unit slope depletion on the well test response [14]. Point bars are often of a particular geometry (ca 3 times as long as wide) in which linear flow will not develop. From depletion, reservoir volumes can be determined – and these will be small if detected during a short (i.e. 24hr) production test. There are relationships published between thickness, width and volume – for point bar sandstones. Of course, in some fluvial reservoirs a mixture of channel body boundaries and fault induced boundaries may be present.

6. Considering other very heterogeneous reservoirs – Carbonate and fractured

A few words are warranted of other even more heterogeneous reservoirs – where aspects of the above will be important to note and the effects may be even more dramatic.

Very high heterogeneity. Carbonates often have even larger ranges of permeability for given porosity and this will translate into even higher measures of variability. Sometimes the presence of vugs are not captured in the core plug data – because of their size. This effect is mitigated by the use of whole core samples – but these are also of limited use where very large vugs are present.

Multiple rock types. Carbonates have many more reservoir pore space creation mechanisms – often diagenetic by origin – which adds to the complexity. Dissolution, vugs, stylolites, microporosity, dolomitisation are just a few of the additional geological phenomena/processes, that impact reservoir properties, to look out for in carbonates.

Fractures. Carbonate (and occasionally fluvial) reservoirs are often fractured. Detecting fractures relies on core, image logs and production logs – being carful not to confuse fractures with high permeability matrix elements as discussed above, does require special attention. Fractures are rarely sampled in core plugs – but where they are often stand out as high permeability, low porosity anomalies.

Well testing considerations. Identification of fractures and boundaries – natural or artificial - from highly heterogeneous reservoirs might be misleading. Complex double matrix porosity considerations with lateral and vertical cross flow effects might be confused with double porosity interpretations. Negative skin is not necessarily a diagnostic signature of a fractured reservoir. Geoskin can result from presence of high-permeability 'pseudo-channels' which are present in make fluvial reservoirs.

7. Effect of heterogeneity on oil recovery

Poor areal and vertical sweep leads to poor oil recovery from a reservoir [15]. Fluvial reservoirs with disconnected channels or partially- and variably-connected, vertical and laterally aggrading sand bodies (the net: gross and the lateral and vertical architectural stacking patterns are critical in this respect) will have very variable flow paths through the system. Rarely will the sweep be uniform –more likely to be fingering, bypassing and dispersive – leading to high remaining mobile oil [16, 1]. Cross-flow is a problem for gas and water flooding as there is little to counteract the effects of gravity and the is often the reason why WAG works well in high net:gross fluvial reservoirs.

Low net to gross fluvial reservoirs require something very different as connectivity is the major challenge and infill drilling may be the answer. Where there is good sand continuity the presence of cross-bedding might impact the capillary trapping of remaining oil.

There is no doubt that fluvial reservoirs are complex and that finding the right engineering solution will be a painstaking and demanding task. Gravity and capillary forces in the reservoir and the viscous-dominated issues in the connectivity of the reservoir to the producing wells have all to be overcome.

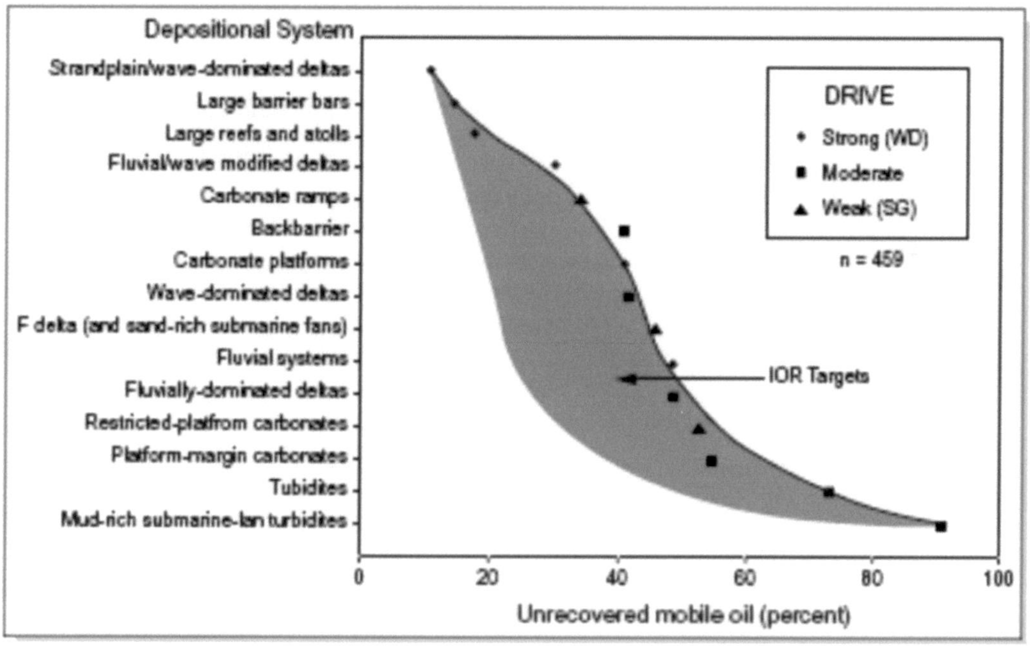

Figure 10. Potential IOR targets in fluvial systems where high amounts of unrecovered oil remain [from 16 and adapted in 1]. A better understanding of the connectivity issues and potential habitat of the unrecovered oil in fluvial reservoirs is a multidisciplinary, geoengineering challenge.

8. Conclusions

All petroleum reservoirs are certainly not of fluvial origin. However, fluvial reservoirs are good reservoirs in which to study the impact of reservoir and that's why I chose them to illustrate this chapter. The reader will have to extrapolate their learnings to other reservoir systems. Fluvial reservoirs are also good reservoirs in which to demonstrate, and to understand the relevance of, close integration between the disciplines. Their understanding and development benefits from such a close integration of geoscience and engineering technology. Through gaining this understanding the reader and industry will doubtless develop improved performance capabilities and thereby engineer higher recovery in these (and other such complex) systems.

Author details

Patrick W. M. Corbett
Institute of Petroleum Engineering, Heriot-Watt University, Edinburgh, UK

Institute of Geosciences, Universidade Federal do Rio de Janeiro, Brazil

Acknowledgement

The author acknowledges the contributions by many students over the years to his understanding and many of those are co-authors in the references. Patrick also acknowledges 17 years of funding from Elf and Total over which period the work on fluvial reservoirs was a major effort.

Symbols

FZI	Flow Zone Indicator
GHE	Global Hydraulic Elements
IOR	Improved Oil Recovery
k_v	Vertical permeability
k_h	Horizontal Permeability
k_x,k_y	Permeability in orthogonal horizontal directions
LP	Lorenz Plot
mD	Milledarcy
MLP	Modified Lorenz Plot
RQI	Reservoir Quality Index
RRT	Reservoir Rock Type
S.D.	Standard Deviation
SG	Solution Gas Drive Mechanism
V_{DP}	Dykstra-Parsons Coefficient
WAG	Water Alternating Gas
WD	Water Drive Mechanism

9. References

[1] Corbett, P.W.M., 2009, *Petroleum Geoengineering: Integration of Static and Dynamic Models*, SEG/EAGE Distinguished Instructor Series, 12, SEG, 100p. ISBN 978-1-56080-153-5

[2] Davies, D. K., Williams, B. P. J. & Vessell, R. K.: "Models for meandering and braided fluvial reservoirs with examples from the Travis Peak Formation, East Texas", SPE 24692, 1992.

[3] Corbett, P.W.M., Zheng, S.Y., Pinisetti, M., Mesmari, A., and Stewart, G., 1998; The integration of geology and well testing for improved fluvial reservoir characterisation, SPE 48880, presented at SPE International Conference and Exhibition, Bejing, China, 2-6 Nov

[4] Cosentino, L., 2001, *Integrated Reservoir Studies*, Editions Technip, Paris, 310p.

[5] Brayshaw, A.C., Davies, R., and Corbett, P.W.M., 1996, Depositional controls on primary permeability and porosity at the bedform scale in fluvial reservoir sandstones, *Advances in fluvial dynamics and stratigraphy*, P.A.Carling and M. Dawson (Eds.), John Wiley and Sons, Chichester, 373-394.

[6] Flint, S.S., and Bryant, I.D., 1993, Geological Modelling of Hydrocarbon Reservoirs and Outcrop Analogues, International Association of Sedimentologists, Wiley, ISBN 9780632033928, Online ISBN 9781444303957

[7] CIPR, University of Bergen, Norway, accessed 19May 2012, http://www.cipr.uni.no/projects.aspx?projecttype=12&project=86

[8] Corbett, P.W.M., and Jensen, J.L., 1992. Estimating the mean permeability: How many measurements do you need? *First Break*, 10, p89-94

[9] Corbett, P.W.M., and Mousa, N., 2010, Petrotype-based sampling to improved understanding of the variation of Saturation Exponent, Nubian Sandstone Formation, Sirt Basin, Libya, *Petrophysics*, 51 (4), 264-270

[10] Ellabad, Y., Corbett, P.W.M., and Straub,R., 2001, Hydraulic Units approach conditioned by well testing for better permeability modelling in a North Africa oil field, SCA2001-50, Murrayfield, 17-19 September, 2001.

[11] Corbett, P.W.M., Ellabad, Y., Egert, K., and Zheng, S.Y., 2005, The geochoke test response in a catalogue of systematic geotype well test responses, SPE 93992, presented at Europec, Madrid, June

[12] Morton, K., Thomas, S., Corbett, P.W.M., and Davies, D., 2002, Detailed analysis of probe permeameter and vertical interference test permeability measurements in a heterogeneous reservoir, *Petroleum Geoscience*, 8, 209-216.

[13] Corbett, P.W.M.,Hamdi, H.,and Gurev, H., 2012, Layered Reservoirs with Internal Crossflow: A Well-Connected Family of Well-Test Pressure Transient Responses, *Petroleum Geoscience*, v18, 219-229.

[14] De Rooij, M., Corbett, P.W.M., and Barens, L., 2002, Point Bar geometry, connectivity and well test signatures, *First Break*, 20, 755-763

[15] Arnold, R., Burnett, D.B., Elphick, J., Freeley III, T.J., Galbrun, M., Hightower, M., Jiang, Z., Khan, M., Lavery, M., Luffey, F., Verbeek, P., 2004, Managing Water – From waste to resource, *The Technical Review,* Schlumberger, v16, no2, 26-41

[16] Tyler, N., and Finley, R.J., 1991, Architectural controls on the recovery of hydrocarbons from sandstone reservoirs, in Miall, A.D., and Tyler, N., (eds.) *The three dimensional facies architecture of terrigeneous clastic sediments and its implications for hydrocarbon discovery and recovery,* SEPM Concepts in Sedimentology and palaeontology, Tulsa, Ok, 3, 1-5

Section 4

Core Analyses

Chapter 9

Digital Rock Physics for Fast and Accurate Special Core Analysis in Carbonates

Mohammed Zubair Kalam

Additional information is available at the end of the chapter

http://dx.doi.org/10.5772/52949

1. Introduction

Initiatives for increasing hydrocarbon recovery from existing fields include the capability to quickly and accurately conduct reservoir simulations to evaluate different improved oil recovery scenarios. These numerical simulations require input parameters such as relative permeabilities, capillary pressures, and other rock and fluid porosity versus permeability trends. These parameters are typically derived from Special Core Analysis (SCAL) tests. Core analysis laboratories have traditionally provided SCAL through experiments conducted on core plugs. Depending on a number of variables, SCAL experiments can take a year or longer to complete and often are not carried out at reservoir conditions with live reservoir fluids. Digital Rock Physics (DRP) investigates and calculates the physical and fluid flow properties of porous rocks. In this approach, high-resolution images of the rock's pores and mineral grains are obtained and processed, and the rock properties are evaluated by numerical simulation at the pore scale.

Comparisons between the rock properties obtained by DRP studies and those obtained by other means (laboratory SCAL tests, wireline logs, well tests, etc.) are important to validate this new technology and use the results it provides with confidence. This article shares a comparative study of DRP and laboratory SCAL evaluations of carbonate reservoir cores.

This technology is a breakthrough for oil and gas companies that need large volumes of accurate results faster than the current SCAL labs can normally deliver. The oil and gas companies can use this information as input to numerical reservoir simulators, fracture design programs, analytic analysis of PTA, etc. which will improve reserve forecasts, rate forecasts, well placement and completion designs. It can also help with evaluating option for improved oil recovery with sensitivity analysis of various options considering the actual pore scale rock fabric of each reservoir zone. Significant investment savings can also be realized using good DRP derived data compared with the conventional laboratory SCAL tests.

2. Background and summary

The main objective is to provide petrophysical and multiphase flow properties, calculated from 3D digital X-ray micro-tomographic images of the selected reservoir core samples. The simulations have been conducted on sub-samples (micro plugs) and then upscaled to cores plug scale for direct comparison with experimental data. Typically the whole core are imaged on dimensions of 11 – 16.5 cm with a resolution of 500 microns, while the core plugs are imaged from to 2 – 4 cm with resolutions of 12 – 19 microns. The micro plugs have dimensions from 1 – 5 mm with resolutions of 0.3 – 5 microns and with Nano-CT one can look at rocks of 50 – 300 microns with resolutions from 50 – 300 nm. In DRP process, the results of these increasingly smaller and smaller investigations are then integrated by an upscaling, either by steady state or geometric methods. Hence, rock properties are computed from Nano and micro scale to plug scale to a whole core scale.

Absolute permeability can be computed using Lattice-Boltzmann simulations, calculation of Formation Resistivity Factor is based on a solution of the Laplace equation with charge conservation (the equations were solved using a random walk algorithm) and elastic properties were calculated by the finite element method. Primary drainage and waterflood capillary pressure and relative permeabilities are determined from multi-phase flow simulations on the pore network representation of the 3D rock model. Flow simulation input parameters were set according to expected wettability conditions.

The first section outlines the basic DRP based results on reservoir properties determined on complex carbonates from giant Middle East reservoirs and compared with similar measurements performed in SCAL laboratories. Section 3 outlines possible details in the calculation process for each of the many reservoir parameters that can be calculated using DRP, while section 4 illustrates snapshots of multi-phase flow results on complex carbonates from the same giant Middle East reservoirs. Capillary pressure, cementation exponent (m), saturation exponent (n) for both primary drainage and imbibition, water-oil and gas-oil relative permeability and elastic properties of carbonates were calculated from DRP. Very good agreement is obtained between DRP derived properties and available experimental data for the studied data set. The results obtained for porosity, absolute permeability, formation resistivity factor, cementation and saturation exponent are shown in Figure 1. through 5. Calculations of elastic properties have been performed on all reconstructed samples. The elastic parameters V_s and V_p are reported in figure 6 and compared to available literature data.

3. Typical DRP Workflow

3.1. Introduction

In order to meet project objectives the workflow illustrated in Figure 7. was implemented. High resolution (19 μm/voxel) micro-CT images of core plugs were first recorded in order to identify rock types and their distribution within the core plugs, and to select locations for thin sections and micro-plugs.

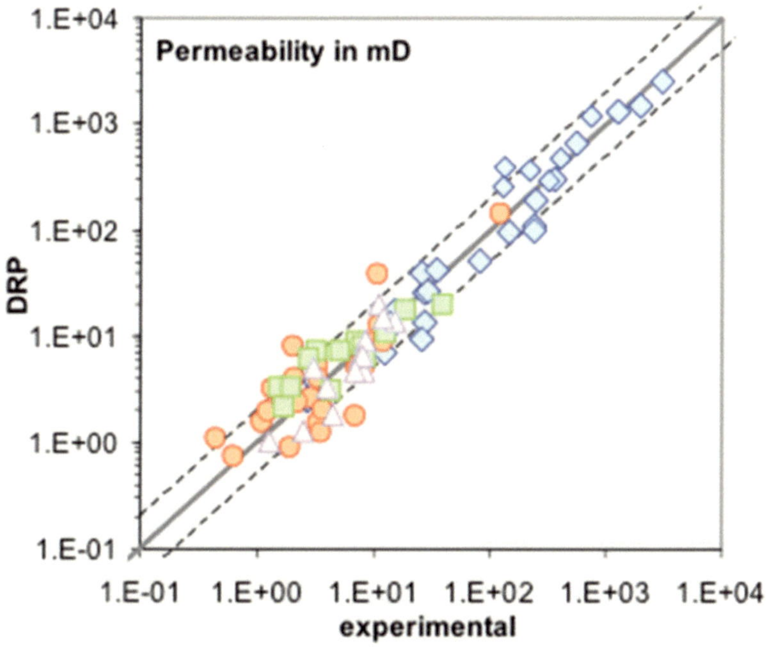

Comparison of experimental and simulated permeability for all samples (straight line is 1:1 and dotted line is factor 2)

Figure 1. Simulated vs. experimental permeability

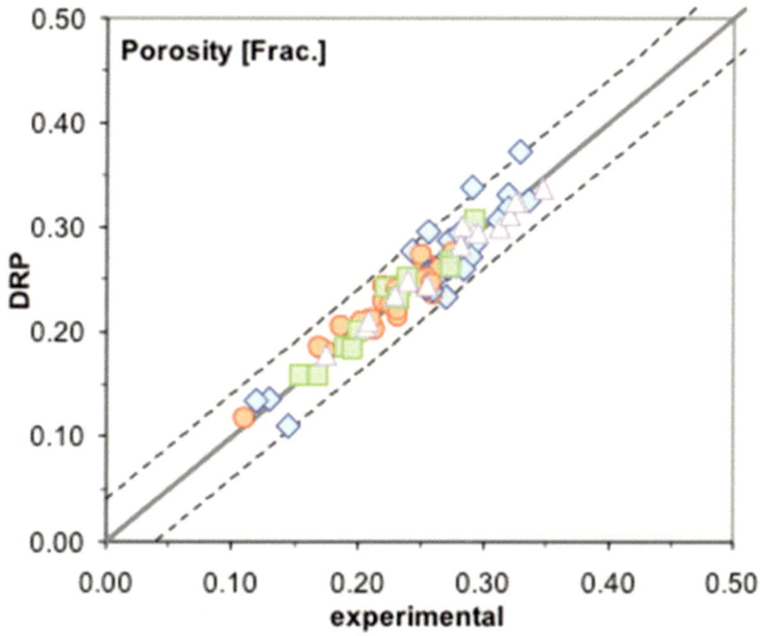

Comparison of experimental and simulated porosity for 100+ samples (straight line is 1:1 and dotted line is +/- 3%)

Figure 2. Simulated vs. experimental porosity

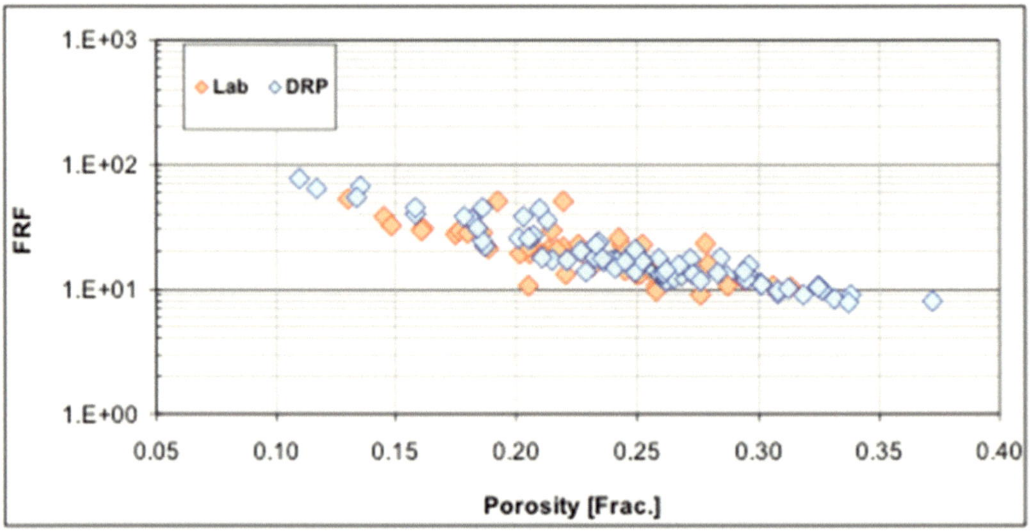

Calculated Formation Resistivity Factor (FRF) as a function of total porosity for the e-Core models compared to available experimental data.

Figure 3. Porosity-FRF correlation

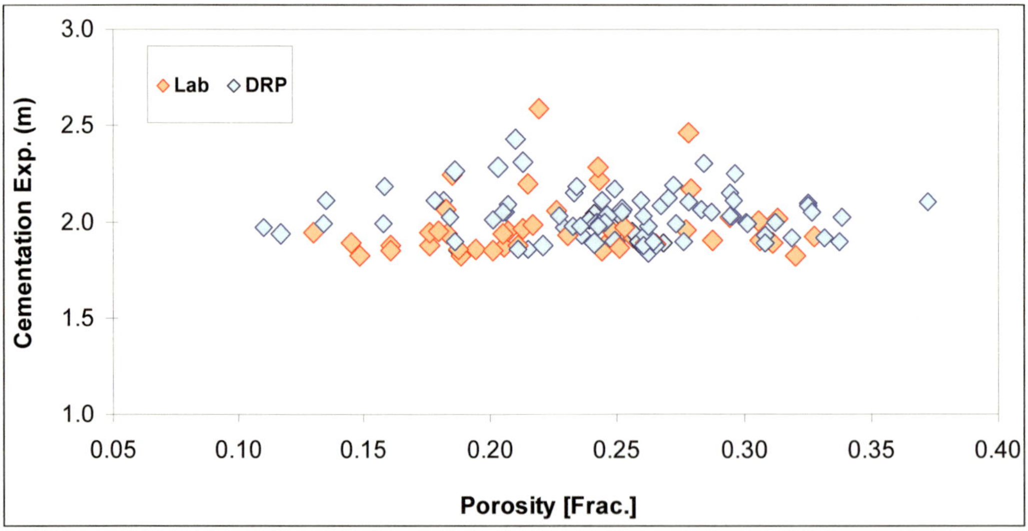

Calculated cementation exponent (m) as a function of total porosity for the e-Core models compared to available experimental data.

Figure 4. Porosity-Cementation exponent correlation

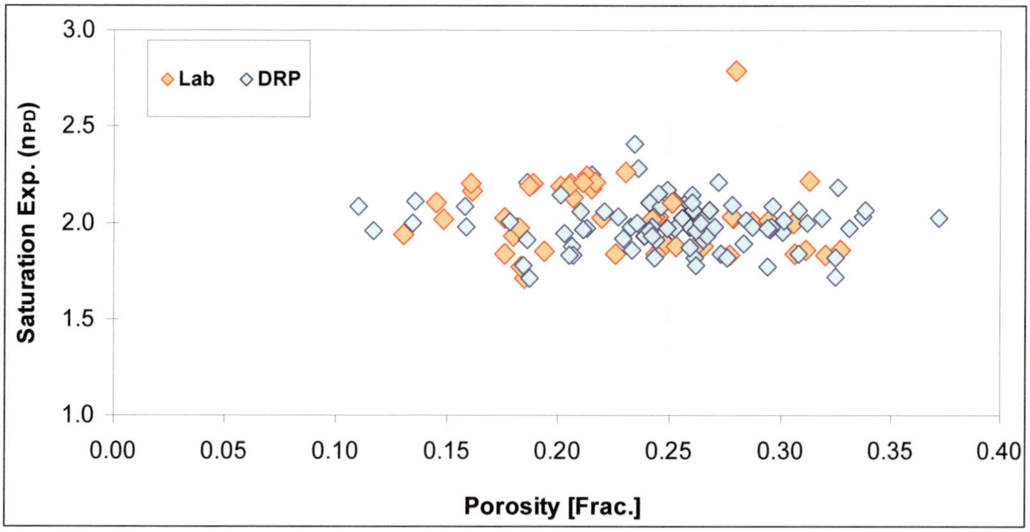

Calculated saturation exponent (n) for primary drainage displacement as a function of total porosity for the e-Core models compared to available experimental data.

Figure 5. Porosity-Saturation exponent correlation

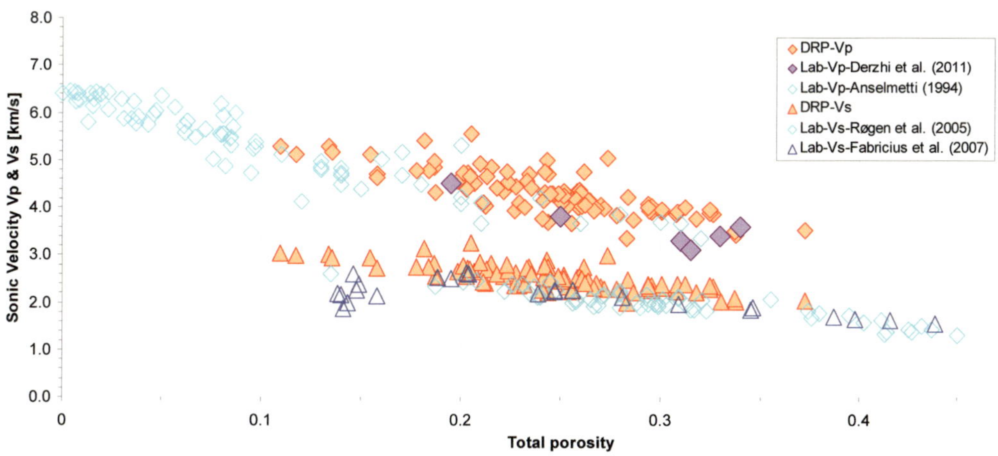

Calculated Vs (open symbols) and Vp (colored symbol) as a function of total porosity compared to available literature data.

Figure 6. Porosity-Vs and Vp correlation

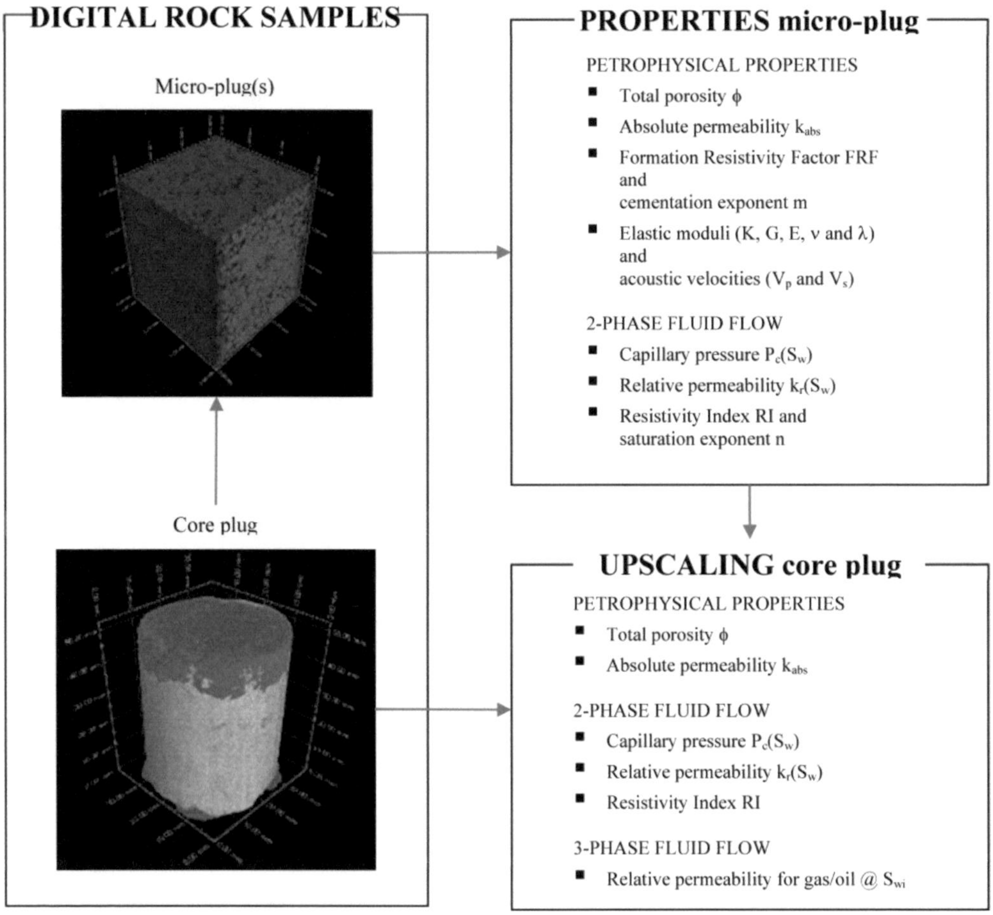

Project workflow and calculated properties.

Figure 7. Digital rock analyses on carbonate samples

Thin sections were then prepared and analysed for each core plug. Micro-plugs were drilled and high resolution (8.0 – 0.3 μm/voxel) micro-CT images were recorded, processed and analysed for each micro-plug, generating 3D digital rock models.

Micro-plug properties have been calculated directly on the digital rock (grid calculations) or on network representations of the pore space (network based calculations):

Grid based reservoir core properties

- Total porosity ϕ
- Absolute permeability k_{abs}
- Formation Resistivity Factor FRF and corresponding cementation exponent m
- Elastic moduli assuming isotropy (bulk modulus K, shear modulus μ, Young's modulus E, Poisson's ratio υ and Lamé's parameter λ) and corresponding acoustic velocities (P-wave velocity V_P and shear-wave velocity V_S)

Network based flow relations

- Capillary pressure P_c as a function of water saturation S_w
- Relative permeability k_r as a function of water saturation S_w
- Resistivity Index RI as a function of water saturation S_w and corresponding saturation exponent n for both primary drainage and imbibition cycle

Capillary pressure and relative permeability curves were established for the following flow processes:

2-phase flow

- Oil/water primary drainage to initial water saturation S_{wi}
- Water/oil imbibition to residual oil saturation S_{orw}

3-phase flow

- Gas/oil drainage at initial water saturation S_{wi}

Resistivity index curves and corresponding saturation exponents were established for water/oil primary drainage and imbibition.

The last step of the workflow is to upscale micro-plug properties (volumes in the mm^3 range) to core plug properties (volumes ranging from ~ 40 to100 cm^3). Rock types were identified and micro-plugs representing these rock types were selected. Each core plug voxel was then assigned to a given rock type or pore space. Corresponding micro-plug and pore fluid properties were used as input to the calculation of upscaled properties.

The following properties were upscaled:

Petrophysical properties

- Total porosity ϕ
- Absolute permeability k_{abs}
- Formation Resistivity Factor FRF
- Cementation Exponent m

Flow properties

- Capillary pressure P_c as a function of water saturation S_w
- Relative permeability k_r as a function of water saturation S_w
- Resistivity Index RI as a function of water saturation S_w

Capillary pressure, relative permeability and resistivity index curves were generated for the micro-plug flow processes listed on previous page.

3.2. Imaging

The principle of microfocus X-ray Computed Tomography (mCT) is based on Beer's law, i.e. the intensity of X-rays is attenuated when passing through physical objects. The attenuated

X-rays are captured by a detector to compose a projection image. A series of projection images from different angles (0 to 360°) are collected by rotating the object around its axis. These projection images are processed to generate 2D mCT slices, which subsequently are input date to construct 3D images of the objects. Each volume-element in these 3D images corresponds to one voxel. The voxel value of the recorded image is proportional to the attenuation coefficient, which is mainly a function of the density and the effective atomic number Z of the constituents of the object (Alvarez and Macovski, 1976; Pullan et al., 1981).

3.3. Thin section preparation, sub-sampling

The petrographic study of thin sections cut from the core plugs is an important part of the workflow. It allows describing and examining the heterogeneity of the core plugs on the micro- to macro-scale. Important features are the identification of micro-facies, the microscopic fabric in each facies type, and the mineralogy of the rock - including diagenetic features (cementation and secondary porosity). The first step in studying the facies associations is the analysis of the initial core plug mCT images (at 19 micron/voxel resolution) to identify the spatial distribution of the various facies, which is the basis for selecting where to cut the slab for thin section preparation, and also to select sites for drilling sub-plugs.

All samples are classified according to Dunham (1962) and Embry & Klovan (1971) limestone Classification. Several rock types are described with special attention to the grain types in terms of size, shape and sorting, because these parameters are influencing the flow properties. One of the main purposes of the microscopic analysis is to describe the porosity of the sample. The different pore types are identified using the Choquette and Pray (1970) porosity classification scheme. Furthermore, the high resolution images of the thin sections are used to distinguish between carbonate cements and micrite, in addition to identifying the main minerals (calcite or dolomite). Drilling sites for micro-plugs are also decided based on the characteristic microfacies types of the sample. The mCT scanning resolution of the micro plugs is chosen in accordance with the microscopic analysis, with particular emphasis on micrite, pore sizes and cementation features.

3.4. Segmentation

3.4.1. Step 1: Cropping and filtering

Scanned mCT images of plugs are cropped into the largest possible undisturbed rectangular volume avoiding cracks, uneven edges and gradients towards the outer surface of the sample. Cropped images are scaled to a voxel size representing approximately the size of the corresponding micro-plug images. A noise reduction filter is applied.

Scanned images of micro-plugs are cropped to a volume of 1200^3 grid cells to avoid gradients at the side of the image and to allow further processing (size limitations of applied software).

3.4.2. Step 2: Histogram Analysis

Each image element carries an 8-bit signal (256 grey values) corresponding to the X-ray attenuation experienced within its volume. A voxel completely filled by empty pore space (air) data is approaching black (0, ρ_{air} = 0.0012 g/cm³), while a voxel completely filled by high density mineral such as pyrite is approaching white (255, ρ_{pyrite} = 4.8 – 5.0 g/cm³). Voxels completely filled by minerals with lower densities, or voxels filled by various proportions of minerals and/or pore space cover the entire grey scale range from black to white. The grey value is consequently not determining the contents of the voxels uniquely; it may represent a single mineral, two or more minerals or porosity and minerals.

An example of a grey value histogram is shown in Figure 8. The histogram itself is shown as blue diamonds and is displayed on logarithmic scale (to the left). The other two data sets (pink and yellow) show the first and second order differentials of the grey value counts.

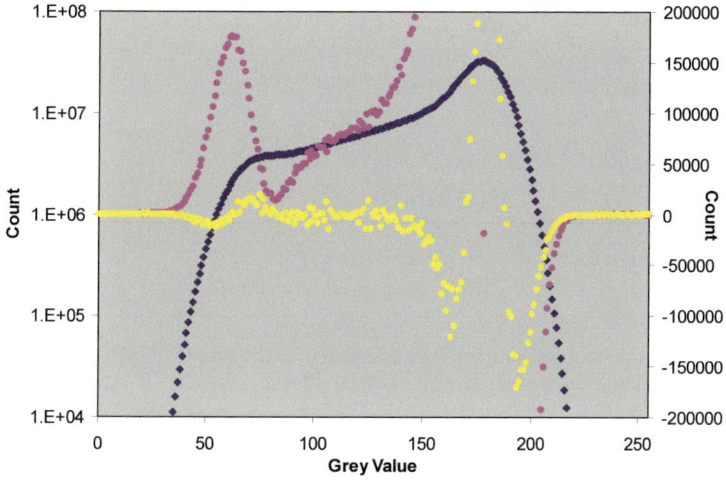

Figure 8. Example grey value histogram.

The following porosities can then be calculated:

$$\phi_{vis} = \sum_{n=o}^{GV_1-1} x_n \tag{1}$$

$$\phi_{\mu} = \sum_{n=GV_1;m=1}^{GV_2} x_2\left(1-\frac{m}{GV_2-GV_1}\right) \tag{2}$$

$$\phi_{res} = \phi_{vis} + \phi_{\mu} \tag{3}$$

where x denotes the fraction of the entire image of the grey value and n and m are counting variables in a grey value range (n is grey value specific, m positive integer to distribute micro-porosity evenly).

This analysis underestimates porosity and permeability. Therefore, a practical threshold needs to be found to decide up until which micro-porosity value voxels are considered pore. For this purpose, the grey value (GV_P) is found where:

$$\sum_{n=0}^{GV_p} x_n = f_{res} \tag{4}$$

This is illustrated in Figure 9. Thus, the total porosity of the image is calculated according to:

$$\phi_{tot} = \phi res + \sum_{n=GV_p+1;\, m=GV_2-GV_p+1}^{GV_2} x_n \left(1 - \frac{m}{GV_2 - GV_1}\right) \tag{5}$$

where the second term gives the amount of micro-porosity extracted from the image. Note that the micro-porosity values for individual grey values are the same in eq. 5 as in eq. 2.

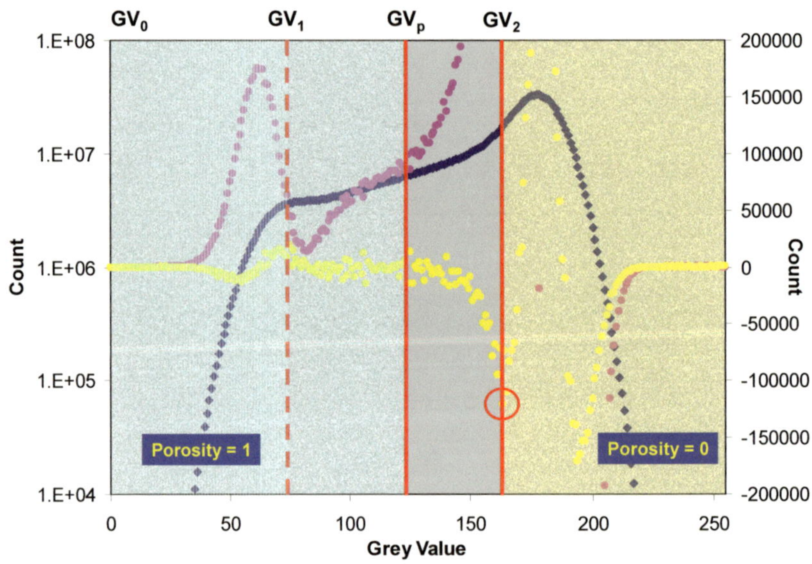

Figure 9. Practical pore threshold.

The grey values GV_P and GV_2 give the segmentation thresholds for the micro-plug images. Three images are prepared according to this segmentation method:

1. representing ϕ_{res} and matrix (including remaining micro-porosity)
2. representing ϕ_{res}, micro-porosity and matrix
3. representing a coarser image where ϕ_{res} and matrix are represented as one phase and micro-porosity is segmented into 6 different porosity ranges

Image 1 is segmented into a pore network representation (see Bakke and Øren, 1997, Øren and Bakke, 2002, 2003) which is used to calculate capillary pressure, relative permeability, resistivity index, and saturation exponent. Image 2 is used to calculate porosity, permeability, formation resistivity factor, and cementation exponent. Image 3 is used in a steady-state upscaling routine to include properties calculated for generic models for micritic material representing the micro-porosity.

The scaled image of the whole core plug is segmented in the same way to identify ϕ_{res} as vugs. However, the remaining matrix of the core plug is segmented into pure solid and 1-3 porosity classes according to the segmentation of the respective micro-plugs extracted from the core plug.

3.5. Calculations/Simulations

3.5.1. Single-phase properties

3.5.1.1. Porosity

Three different porosities are reported: total porosity (ϕ_{tot}), micro-porosity (ϕ_{μ}) and percolating porosity (ϕ_{perc}). ϕ_{perc} is the fraction of ϕ_{res} (see above) that is available to flow, i.e. accessible from any side of the micro-plug image, expressed as a fraction of the whole micro-plug volume. The fraction of ϕ_{res} that is not available for flow is isolated porosity (ϕ_{iso}). ϕ_{μ} is all porosity present in voxels between GV_P and GV_2 expressed as a fraction of the whole micro-plug volume. ϕ_{perc} is only reported for the micro-plug images. ϕ_{tot} and ϕ_{μ} are reported for both the micro-plugs and the whole core plug. Some micro-plug images with the highest resolution can be considered entirely as micro-porosity. The following relations hold for porosities of micro-plug and core plug images:

$$\begin{aligned} \phi_{tot,coreplug} &= \phi_{res,vugs} + f_{\mu-plug,A}\phi_{tot,A} + f_{\mu-plug,B}\phi_{tot,B} + f_{\mu-plug,C}\phi_{tot,C} \\ \phi_{tot,\mu-plug} &= \phi_{\mu} + \phi_{res} \\ \phi_{res,\mu-plug} &= \phi_{perc} + \phi_{iso} \end{aligned} \quad (6)$$

where f is fraction.

3.5.1.2. Absolute permeability

A Lattice-Boltzmann method is applied to solve Stokes' equation in the uniform grid model. Flow is driven either by a constant body force or a constant pressure gradient through the model. Permeability is calculated in three orthogonal directions separately. Sides perpendicular to flow directions are closed during each directional calculation (no-flow boundary conditions). In this study, absolute permeability is calculated using a constant body force because this setting delivers more accurate results when model resolution is sufficient and porosity is relatively high. Averages of three directional calculations are reported for each model realization. For further details see Øren and Bakke (2002).

3.5.1.3. Formation resistivity factor

The steady state electrical conductivity, or formation resistivity factor (F), of a brine saturated rock is governed by the Laplace equation

$$\begin{aligned} &\nabla \bullet \mathbf{J}=0 \\ &\mathbf{J}=\sigma_w \nabla \Phi \end{aligned} \tag{7}$$

subject to the boundary condition $\nabla\Phi \bullet \mathrm{n}=0$ on the solid walls (i.e. insulating walls). Here **J** is the electrical current, σ_w is the electrical conductivity of the fluid that fills the pore space, Φ is the potential or voltage and **n** is the unit vector normal to the solid wall. Numerical solutions of the Laplace equation are obtained by a random walk algorithm or by a finite difference method (*Øren and Bakke*, 2002).

The effective directional conductivities σ_i, $i = x, y, z$ are computed by applying a potential gradient across the sample in *i*-direction. The directional formation resistivity factor F_i is the inverse of the effective electrical conductivity $F_i = \sigma_i / \sigma_w$. We define the average formation resistivity factor F as the harmonic mean of direction dependent formation factors. The cementation exponent m is calculated from F and the sample porosity using Archie's law $F = \phi^{-m}$.

Formation factor and cementation exponent reported were approximately 10-15% greater than experimental data. The F and m values reported in the previous version were calculated as described above, i.e. by assuming that only the resolved porosity contributes to the conductivity. Micro porosity present in the micrite phase was thus treated as insulating solid. In the second version, we accounted for the conductivity of the micrite phase by assigning a finite conductivity σ_{mic} to micrite voxels using Archie's law $\sigma_\mu = \sigma_w (\phi\, \sigma_\mu)^m$, where ϕ_μ and m are the micrite porosity and the cementation exponent of the micrite phase.

3.5.1.4. Elastic moduli

Pore-scale modeling of elastic properties

Numerical code

The finite element method described by Garboczi and Day (1995) has been implemented for calculation of elastic properties. The method uses a variational formulation of the linear elastic equations and finds the solution by minimizing the elastic energy using a fast conjugate-gradient method. The results are valid for quasi-static conditions or at frequencies which are sufficiently low such that the included pore pressures are in equilibrium throughout the pore space (Arns *et al.*, 2002).

The effective bulk and shear moduli are computed assuming isotropic linear elastic behavior. The V_P and V_S are subsequently calculated using the simulated effective elastic moduli and the effective density according to:

$$V_p = \sqrt{\frac{K + \frac{4}{3}\mu}{\rho}} \tag{8}$$

$$V_s = \sqrt{\frac{\mu}{\rho}} \tag{9}$$

Inputs to the calculations are:

- A three-dimensional representation of the rock microstructure (a digital rock sample)
- Density ρ and elastic properties (K and μ) of each mineral composing the rock matrix
- Density ρ and elastic properties (K and μ) of the fluid present in the pore space

Pore-scale versus experimental data

Scale and sample selection

Acoustic measurements are generally performed on samples with volumes ranging from 40 to 100 cm^3. Digital rock samples are significantly smaller, generally in the mm^3 range. Care must therefore be taken when laboratory measurements and pore-scale derived properties are compared, due to the scale difference. Laboratory samples and plugs for μCT imaging are not only of different volumes, they are also generally sampled at different locations. Due to the spatial variability, it is recommended to compare trends when laboratory measurements are compared with pore-scale derived properties. This is illustrated in Figure 10 for the P-wave velocity. A digital sample with 12.9% porosity has been made. However, samples tested in the laboratory do not have porosities close to this value. By sub-sampling the original digital sample, a relative broad porosity range is obtained (from 9.9 to 16.3%). The pore-scale derived velocity – porosity trend is now overlapping with the measurements performed in the laboratory, and a comparison can be made.

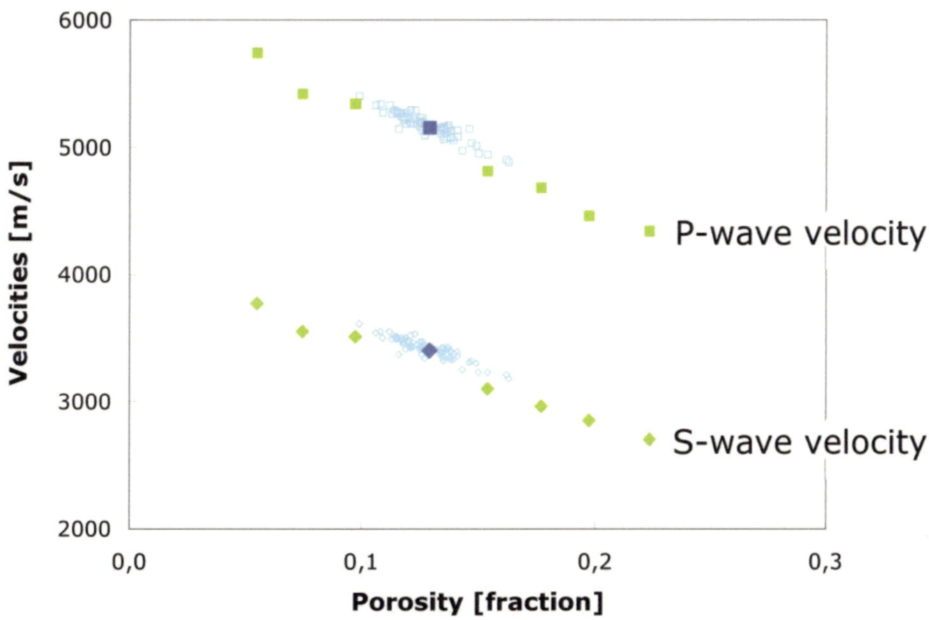

Pore-scale derived P- and S-wave velocities (dark and light blue symbols) are compared with laboratory measurements (green filled symbols). Dark blue points represent a large digital "mother" sample, while the light blue points represent sub-samples of the original "mother" sample.

Figure 10. Fontainebleau sandstone.

Stress

Cracks may reduce the acoustic velocities significantly. Sub-resolution cracks are not incorporated in the processed mCT images and the corresponding pore-scale derived velocities are therefore overestimated for materials containing such cracks.

Cracks are closed during loading. Acoustic measurements performed at elevated stress levels are consequently expected to approach pore-scale derived velocities. Derzhi and Kalam (2011) compared acoustic measurements at different stress levels with pore-scale derived velocities. Their results are shown in Figure 11. Note that this assumes that the pore space and rock framework deforms without large micro-structural changes such as pore collapse, grain rotation and grain crushing.

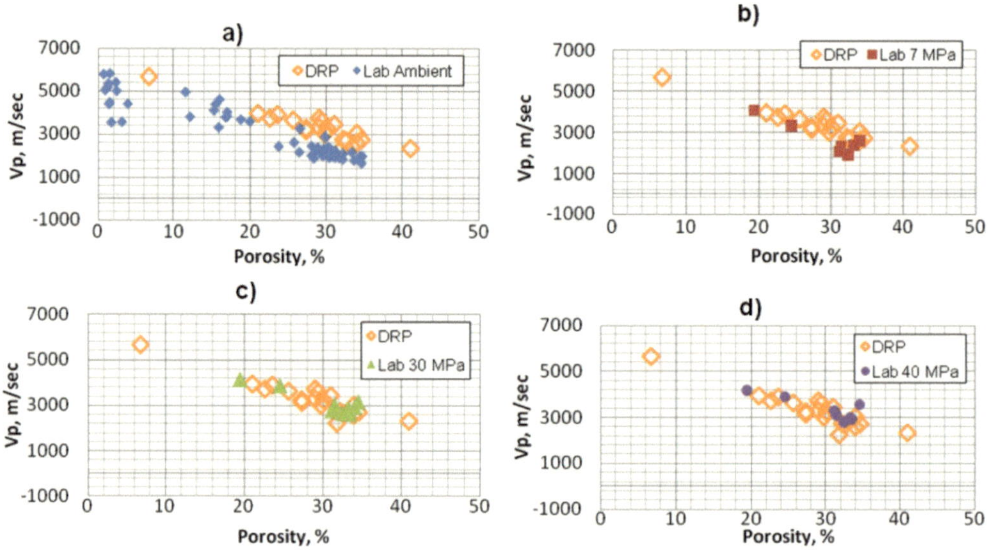

P-wave velocity as a function of porosity at different stress levels: a) at ambient conditions, b) at 7 MPa, c) at 30 MPa and d) at 40 MPa effective hydrostatic stress. Pore-scale derived values are denoted DRP and shown as open orange triangles. Laboratory measurements are shown as filled symbols; from Derzhi *et al.* (2011).

Figure 11. P-wave velocity and stress for carbonate samples.

Frequency

Pore-scale derived elastic properties represent properties in the low frequency limit (f $\rightarrow 0$). Measurements in the laboratory are generally performed in the ultrasonic range (1 Hz to 4 kHz). Acoustic velocities are independent of frequency for dry materials, while an increase with increasing frequency has been observed for fluid saturated rocks. Pore-scale derived properties are therefore expected to be comparable to laboratory measurements for dry rocks – and lower for fluid saturated rocks.

3.5.1.5. NMR

NMR is simulated as a diffusion process using a random walk algorithm to solve the diffusion equations (Øren, Antonsen, Rueslåtten and Bakke, 2002). The T2 responses at S_w =

1.0 were simulated on one sample from the field A. The results are shown as T2 decay and T2 distribution curves in figures 14 and 15. The surface relaxation strength was kept at 1.65×10^{-5} m/sec for all minerals. The inter echo time was 200μsec at a background magnetic field gradient of 0.2 G/m, the bulk water T2 was 0.3 sec, and the diffusion constant was 2×10^{-9} m^2/sec. The decay curves are transformed into T2 distributions using 50 exponential functions (assuming the same has been done for the lab data).

3.5.2. *Multi-phase flow simulations*

The reconstructed rock models were simplified into pore network models. Crucial geometrical and topological properties were retained, while the data volume was reduced to allow timely computation (Øren and Bakke, 2003). In pore network modelling, local capillary equilibrium and the Young–Laplace equation are used to determine multiphase fluid configurations for any pressure difference between phases for pores of different shape and with different fluid/solid contact angles. The pressure in one of the phases is allowed to increase and a succession of equilibrium fluid configurations are computed in the network. Then, empirical expressions for the hydraulic conductance of each phase in each pore and throat are used to define the flow of each phase in terms of pressure differences between pores. Conservation of mass is invoked to find the pressure throughout the network, assuming that all the fluid interfaces are fixed in place. From this the relationship between flow rate and pressure gradient can be found and hence macroscopic properties, such as absolute and relative permeabilities, can be determined (Øren and Bakke, 2002).

The following oil-water displacements were simulated:

Primary drainage
Imbibition at a given wettability preference

For each displacement process, capillary pressure and relative permeability curves were calculated. Resistivity index with the corresponding n-exponent was calculated after primary drainage.

Each saturation and relative permeability value corresponds to a capillary equilibrium state. In all the simulations, it is assumed that capillary forces dominate. This is a good approximation for capillary numbers $N_{ca} < 1e^{-6}$. This, however, does not necessarily mean that it is the most efficient displacement possible. In certain cases, viscous and/or gravity forces can dominate and result in higher (or lower) displacement or sweep efficiency.

3.5.2.1. *Relative permeability*

Simulated relative permeability data are fitted to the empirical LET expression (Lomeland, Ebeltoft and Thomas, 2005). For water/oil displacements the LET equations become:

$$k_{ro} = k_{ro}(S_{wi})\frac{(S_{on})^{L_o}}{(S_{on})^{Lo} + E_o(1-S_{on})^{T_o}} \tag{10}$$

$$k_{rw} = k_{rw}\left(S_{or}\right)\frac{\left(S_{wn}\right)^{L_w}}{\left(S_{wn}\right)^{L_w} + E_w\left(1-S_{wn}\right)^{T_o}} \tag{11}$$

$$S_{wn} = \frac{S_w - S_{wi}}{1-S_{wi}-S_{oy}}, \quad S_{on} = 1 - S_{wn} \tag{12}$$

where k_{ro} and k_{rw} are the relative permeability of oil and water, respectively. The L_i's, E_i's, and T_i's are the LET fitting parameters, where i is either oil (o) or water (w). $k_{ro}(S_{wi})$, $k_{rw}(S_{or})$, S_{wi} and S_{or} are determined from the computed results and the optimised values of the fitting parameters were determined using a simulated annealing algorithm.

3.5.2.2. Capillary pressure

A detailed account of the methods used to calculate capillary pressure as a function of S_w during primary drainage and waterflooding invasion sequences is given in Øren et al. (1998). Fluid injection is simulated from one side of the model (usually x-direction). Thus, the entry pressure is a function of the pore sizes present in the inlet. In a mercury injection capillary pressure simulation, fluid is allowed to enter the model from all sides. In that case, entry pressures are much lower.

The calculated capillary pressures can be expressed in terms of the dimensionless Leverett *J*-function:

$$J\left(S_w\right) = \frac{P_{c,ow}}{\sigma_{ow}}\sqrt{\frac{k}{\iota}} \tag{13}$$

where k and ϕ are the permeability and total porosity, respectively. $P_{c,ow}$ is the oil-water capillary pressure and σ_{ow} the oil-water interfacial tension. The Leverett *J*-function results have been fitted to the empirical Skjæveland correlation (Skjæveland et al, 2000):

$$J_{(S_w)} = \frac{c_1}{\left[\frac{\left(S_w - S_{wi}\right)}{1-S_{wi}}\right]} + \frac{c_2}{\left[\frac{\left(1-S_w-S_{or}\right)}{\left(1-S_{or}\right)}\right]} \tag{14}$$

where c_1, c_2, a_1 and a_2 are curve fitting parameters. S_w is the water saturation, S_{wi} initial water saturation and S_{or} residual oil saturation. S_w, S_{wi} and S_{or} is determined by the simulations. Here, results are reported both as J-function and capillary pressure.

3.5.2.3. Water saturation

All reported S_w is total S_w, i.e. including water in microporosity and isolated pores. It should be noted that S_{wi} is strongly dependent on capillary pressure. Thus, any comparison with laboratory data should be done at the same capillary pressure. Any isolated pore volume and any μ-porosity cannot be invaded and, thus, contributes to S_{wi}. Therefore, the irreducible water saturation is given by:

$$S_{wi} = \frac{\phi_{isolated} + \phi_{\mu}}{\phi_{tot}} + S_{w,corner} \tag{15}$$

$$S_{w,corner} = \sum_{1}^{i}\sum_{1}^{n}\left(\frac{\sigma_{ow}}{Pc_{ow,\max}}\right)^2\left(\cos\ \theta_{ow}^r \frac{\cos\left(\theta_{ow}^r + \beta\right)}{\sin\beta} - \frac{\pi}{2} + \beta + \theta_{ow}^r\right) \tag{16}$$

where Φ is porosity (with suffixes denoting isolated, total and microporosity), i is the number of connected pores in the network, n the number of corners per pore (3 or 4), σ interfacial tension, Pc capillary pressure, θ contact angle and β the corner half angle. Note that initial water saturation for waterflooding depends on Pc and can be given as an input to the simulation if needed.

3.5.2.4. Resistivity Index and saturation exponent, n

The resistivity index is calculated from capillary dominated two-phase flow simulations on the pore network representation of the 3D rock image. The basis for simulating capillary dominated displacements is the correct distribution of the fluids in the pore space. For two-phase flow, the equilibrium fluid distribution is governed by wettability and capillary pressure and can be found by applying the Young-Laplace equation for any imposed pressure difference between the phases. A clear and comprehensive discussion of all the mathematical details, including the effects of wettability, involved in the simulations can be found in *Øren et al.*, 1998, and *Øren and Bakke*, 2003).

The current I between two connecting nodes i and j in the network is given by Ohm's law

$$I_{ij} = \frac{g_{ij}}{L_{ij}}\left(\Phi_i - \Phi_j\right) \tag{17}$$

where L_{ij} is the spacing between the node centres. The effective conductance g_{ij} is the harmonic mean of the conductances of the throat and the two connecting nodes

$$\frac{L_{ij}}{g_{ij}} = \frac{l_i}{g_i} + \frac{l_t}{g_t} + \frac{l_j}{g_j} \tag{18}$$

where the subscript t denotes the pore throat and the conductance g_t is evaluated at the throat constriction. The effective lengths l_i, l_j, and l_t govern the potential drop associated with the nodes and the throat, respectively. By letting $l_t = \alpha L_{ij}$ and $l_i = l_j = 0.5(1\text{-}\alpha)L_{ij}$, the effective lengths can be calculated from α as

$$\alpha = \frac{2g_t\left(\frac{1}{g_{ij}} - \frac{1}{g_i} - \frac{1}{g_j}\right)}{g_{ij}\left(1 - \frac{g_t}{g_i} - \frac{g_t}{g_j}\right)} \tag{19}$$

The conductance of a pore element k (pore body or throat) is given by $g_k = \sigma_w A_w$, where A_w is the area of the pore element filled with water. Expressions for A_w for different fluid configurations, contact angles, and pore shapes are given in *Øren et al.*, 1998. We impose current conservation at each pore body, which means that

$$\sum_j I_{ij} = 0 \tag{20}$$

where j runs over all the pore throats connected to node i. This gives rise to a set of linear equations for the pore body potentials. The formation resistivity factor of the network is computed by imposing a constant potential gradient across the network and let the system relax using a conjugate gradient method to determine the node potentials. From the potential distribution one may calculate the total current and thus the formation resistivity factor $F = \sigma_0/\sigma_w$, where σ_0 is the conductivity computed at $S_w = 1$.

The resistivity index is computed similarly. At various stages of the displacement (i.e. different S_w values), we compute the current and the resistivity index defined as

$$RI(S_w) = \frac{\sigma_0}{\sigma(S_w)} = S_w^{-n} \tag{21}$$

The n-exponent is determined from a linear regression of the $RI(S_w)$ vs. S_w. curve.

3.5.3. Upscaling of single phase and effective flow properties

Effective properties of the core samples are determined using steady state scale up methods. The CT scan of the core plug is gridded according to the observed geometrical distribution of the different rock types or porosity contributors. Each grid cell is then populated with properties calculated on the pore scale images of the individual rock types. The following properties are assigned to each grid cell; porosity, absolute permeability tensor (k_{xx}, k_{yy}, k_{zz}), directionally dependent m-exponents, capillary pressure curve, relative permeability curve, and n-exponent.

Single phase up-scaling is done by assuming steady state linear flow across the model. The single phase pressure equations are set up assuming material balance and Darcy's law

$$\begin{aligned} &\nabla \bullet (k\nabla P) = O \ \textit{with boundary conditions} \\ &p = P_1 \ \textit{at} \ x = o, p = P_0 \ \textit{at} \ x = L \\ &v \bullet n = 0 \ \textit{at other faces} \end{aligned} \tag{22}$$

The pressure equation is solved using a finite difference formulation. From the solution one can calculate the average velocity and the effective permeability using Darcy's law. By performing the calculations in the three orthogonal directions, we can compute the effective or up-scaled permeability tensor for the core sample. The effective formation resistivity factor is computed in a similar manner by replacing pressure with voltage, flow with current, and permeability with electrical conductivity. The up-scaled *m*-exponent is determined from the effective formation resistivity factor and the sample porosity.

Effective two-phase properties (i.e. capillary pressure, relative permeability, and *n*-exponent) are calculated using two-phase steady state up-scaling methods. We assume that the fluids inside the sample have come to capillary equilibrium. This is a reasonable assumptions for small samples (<30cm) when the flow rate is slow (<1m/day). The main steps in the two-phase up-scaling algorithm are:

1. Select a capillary pressure (P_c) level
2. Using the P_c (S_w) relationship, calculate S_w in each grid cell
3. Calculate the average water saturation using pore volume weighting
4. From the $k_r(S_w)$ curves, calculate k_{rw} and k_{ro} in each grid cell, and then the phase mobilities k_w and k_o by multiplying the relative permeabilities with the absolute permeability for the grid cell
5. Perform two separate single phase steady state simulations, one for the oil and one for the water, to calculate the effective phase permeability
6. Divide the phase permeability with the effective absolute permeability to obtain the effective relative permeability
7. Repeat these steps with different P_c levels to construct the effective relative permeability curves

The resistivity index curve is generated in a similar manner by replacing phase permeability in the water phase calculations with electrical conductivity. The effective *n*-exponent is determined from a linear regression of the effective $RI(S_w)$ vs. S_w. curve.

3.6. Uncertainty

The overall uncertainty in up-scaled properties varies from sample to sample. Uncertainties may be introduced in the following steps:

- **Generation of digitized core models**
  How representative are identified rock types and their corresponding distribution? In other words; how representative are the digitized model of the core samples?
- **Generation of digitized micro-plugs**
  How representative are μCT models of the rock micro-structure? The main uncertainty is related to size (REV) and image segmentation where the spatial distribution of pore-space and rock minerals is set.
- **Calculation of absolute permeability and formation resistivity factor**
  An uncertainty of ± 2% related to the accuracy of pressure solvers.
- **Calculation of elastic properties**
  Uncertainty related to whether all relevant physics are included in the calculations or not.
- **Simulation of two- and three-phase flow**
  Wettability is an input parameter to the simulations. The uncertainty in wettability is the main source of uncertainty for both two- and three-phase flow simulations.

4. DRP applications in multi-phase flow

In this section, we present some novel results of validation of multi-phase flow SCAL results using Digital Rock Physics. The reservoir cores comprise complex carbonates from giant producing reservoirs in Middle East. Figure 12 show the comparisons of water-oil capillary pressure (Pc) measured in a SCAL laboratory at reservoir temperature and net over burden pressure using a Porous Plate and MICP trims from the same cores corrected to the reservoir conditions. The DRP data were acquired from the cores after the tests were completed on the Porous Plate and core thoroughly cleaned for final SCAL reference measurements. Both limestone and dolomite samples show excellent similarity of DRP derived data with the laboratory evaluations. Figure 13 confirms the validity of such measurements on different sets of core samples comprising the same reservoir rock type (RT), provided the rock typing is valid and captures the key formation properties of rock and fluids.

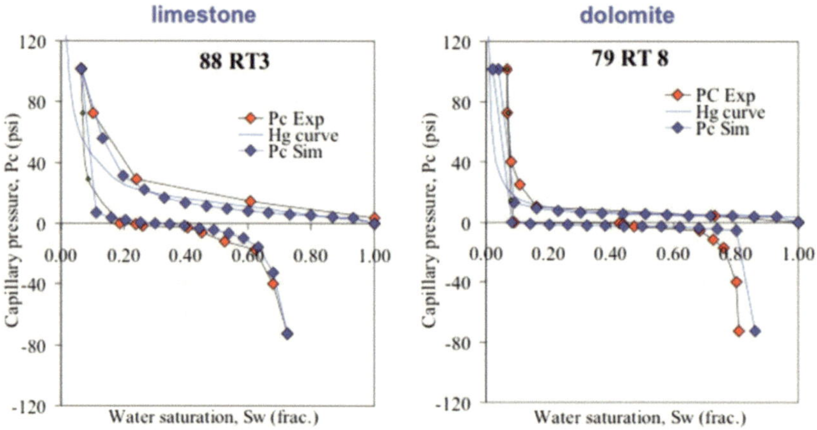

Figure 12. Water-oil Pc (Porous Plate): DRP vs lab on same core sample

Figures 14 and 15 show for the first time in industry that laboratory NMR T2 and MICP measurements done on carbonate rock types can also be captured using DRP based simulations on the same cores with distinctly different pore geometries. The robustness of DRP in capturing NMR T2 based pore bodies and MICP based pore throat distributions have far reaching consequences. This shows that DRP in essence can be used confidently to quantify pore body and pore throat distributions, and therefore the 3D pore geometry is representative of the specific core sample and pore network topology. In using DRP effectively, it is recommended that one compares and validates measured NMR T2 and MICP prior to detailed simulations to quantify various two-phase and three-phase flow properties through such reservoir rocks.

Figures 16 and 17 demonstrate example DRP based validations with respect to water-oil relative permeabilities conducted at full reservoir conditions (reservoir temperature, reservoir pressure and live fluids) on other complex carbonates, including highly permeable vuggy samples. The imbibition displacements were conducted under steady state conditions at SCAL laboratories and QC'ed thoroughly with respect to production, pressure profiles and insitu saturation data, and the corresponding numerically simulated measrements. The

DRP data were acquired on cores comprising each of the composites tested. It is interesting to note that when plug DRP data are compared with composite laboratory measurements there is some scatter and divergence for each reservoir rock type. However, the divergences are significantly minimized when the DRP plugs used are digitally butted to represent the composite used in the laboratory tests. DRP captures the full reservoir condition multi-phase flow data very well, and in some cases even show the experimental artefacts of the SCAL measurements. The validity of tehse tests were confirmed on 14 different reservoir rock types comprising different formations.

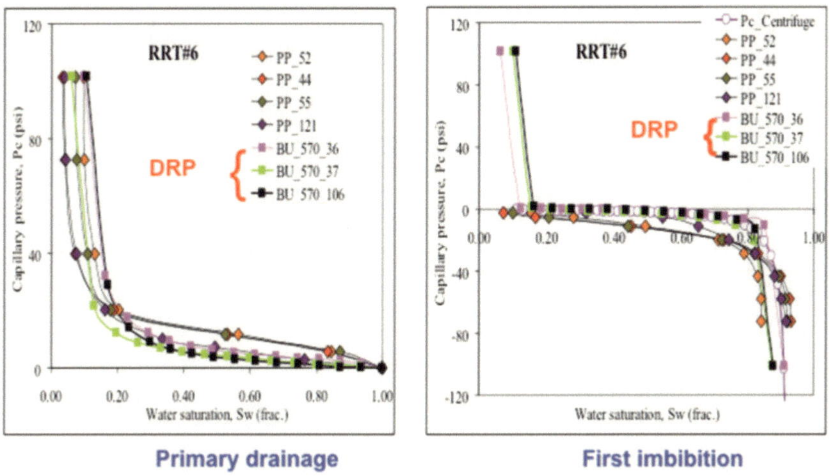

Figure 13. Water-oil Pc (Porous Plate): DRP vs lab in different core samples, but same RRT

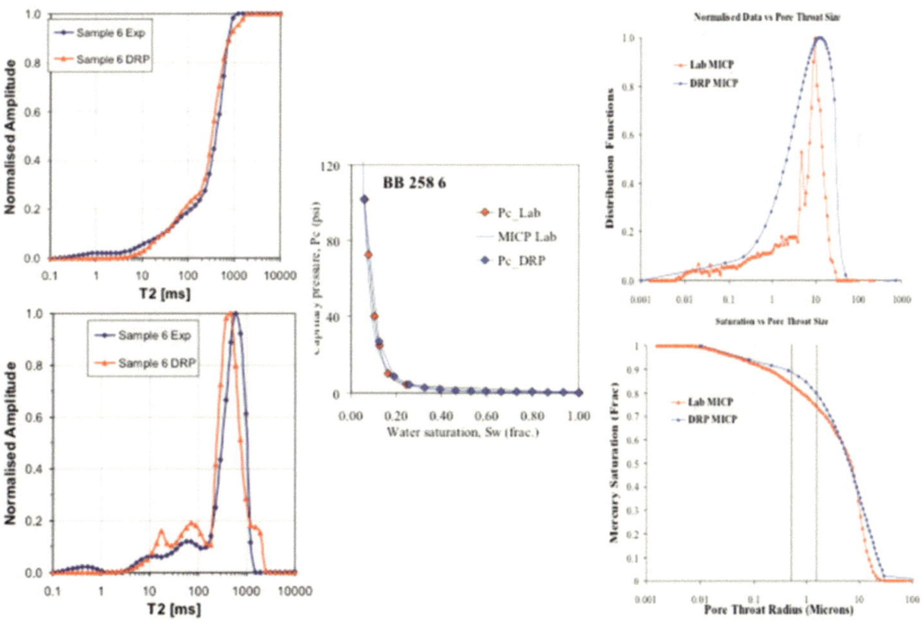

Figure 14. NMR T2 distribution and MICP pore throat distribution, DRP vs Lab – vuggy core

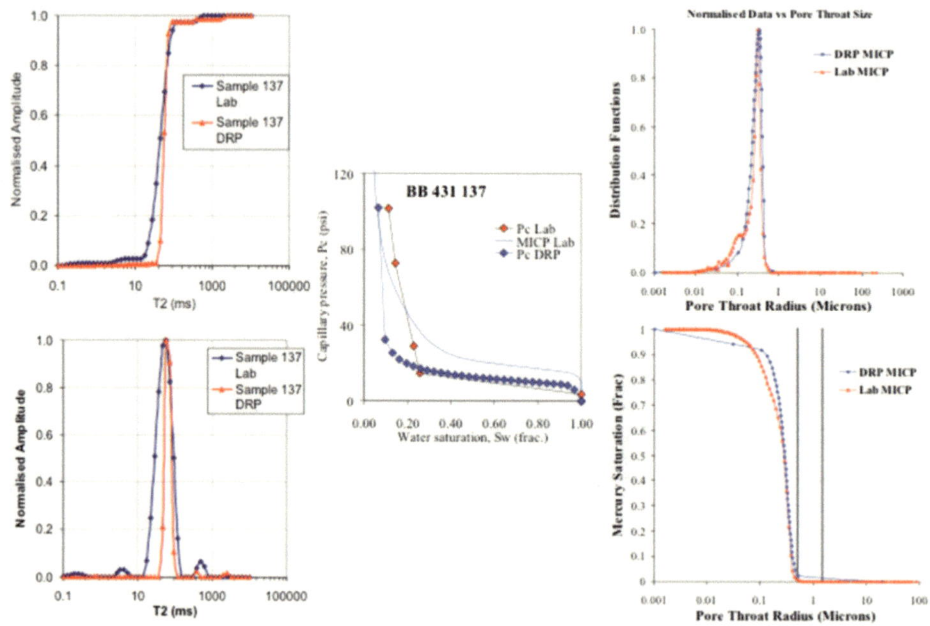

Figure 15. NMR T2 distribution and MICP pore throat distribution, DRP vs Lab – tight core

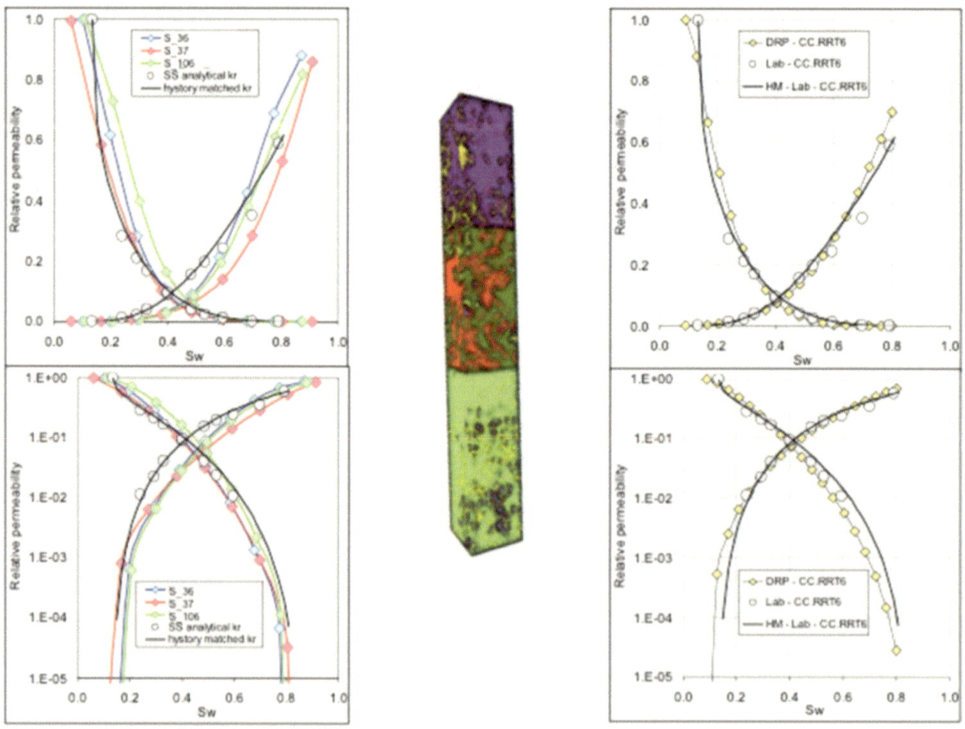

Figure 16. Validating water-oil kr of low permeability composite samples: RRT 6 (10-25 mD)

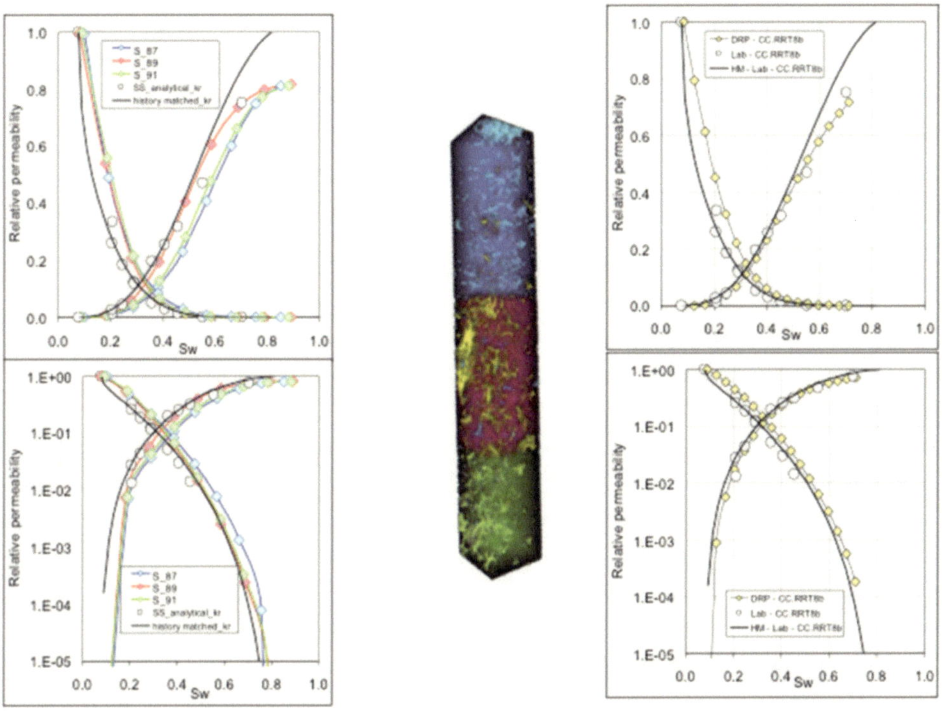

Figure 17. Validating water-oil kr of high permeability composite samples: RRT 8 (350-560 mD)

Author details

Mohammed Zubair Kalam
Abu Dhabi Company for Onshore Oil Operations (ADCO), Abu Dhabi, UAE

Acknowledgement

ADCO and ADNOC Management are acknowledged for their permission to publish these novel Digital Rock Physics based SCAL results.

Numerical Rocks (Norway) is acknowledged for providing the detailed drafts relating to the procedures adopted in example DRP computations and the robust measurements presented in this chapter.

Ingrain Inc (Houston and Abu Dhabi) are acknowledged for introducing the author to the various challenges ahead, and the uncertainties in current DRP based dvelopments.

Digital Core (Australia) is gratefully remembered in first introducing the concept of DRP to Middle East, and involving ADCO in one of the first Joint Industy Projects offered to industry.

iRocks (Beijing and London) are acknowledged for many stimulating discussions relating to the state-of-the-art.

Finally, one must remember the ADCO DRP team for the interest generated and the numerous insights to imaging, segmentation, data evolution and their impact on different Petrophysical, SCAL and elastic properties. I thank my son, AbdAllah, for helping me in getting this draft ready despite the very busy schedules of August 2012.

5. References

Bakke, S., Øren, P.-E., "3-D Pore-Scale Modelling of Sandstones and Flow Simulations in the Pore Networks," *SPE Journal* (1997) 2, 136-149.

Blunt M., Jackson M.D., Piri M., Valvatne P.H., "Detailed physics, predictive capabilities and macroscopic consequences for pore network models of multiphase flow," *Advances in Water Resources* (2002) 25, 1069-1089.

Ghous, A., Knackstedt, M.A., Arns, C.H., Sheppard, A.P., Kumar, R.M., Senden, T.J., Latham, S., Jones, A.C., Averdunk, H. and Pinczewski, W.V., "3-D imaging of reservoir core at multiple scales: Correlations to petrophysical properties and pore-scale fluid distributions," presented at International Petroleum Technology Conference, Kuala Lumpur, Malaysia, 2008, 10 p.

Gomari, K. A. R., Berg, C. F., Mock, A., Øren, P.-E., Petersen, E. B. Jr., Rustad, A. B., Lopez, O., 2011, Electrical and petrophysical properties of siliciclastic reservoir rocks from pore-scale modeling, paper SCA2011-20 presented at the 2011 SCA International Symposium, Austin, Texas.

Grader, A., Kalam, M. Z., Toelke, J., Mu, Y., Derzhi, N., Baldwin, C., Armbruster, M., Al Dayyani, T., Clark, A., Al Yafei, G. B. And Stenger, B., 2010, A comparative study of DRP and laboratory SCAL evaluations of carbonate cores, paper SCA2010-24 presented at the 2010 SCA International Symposium, Halifax, Canada.

Kalam, M.Z., Al Dayyani, T., Grader, A., and Sisk, C., 2011, 'Digital rock physics analysis in complex carbonates', World Oil, May 2011.

Kalam M.Z., Serag S., Bhatti Z., Mock A., Oren P.E., Ravlo V. and Lopez O., SCA2012-03, "Relative Permeability Assessment in a Giant Carbonate Reservoir Using Digital Rock Physics," SCA 2012 International Symposium, Aberdeen, United Kingdom.

Kalam, M.Z., Al-Hammadi, K., Wilson, O.B., Dernaika, M., and Samosir, H., "Importance of Porous Plate Measurements on Carbonates at Pseudo Reservoir Conditions," SCA2006-28, presented at the 2006 SCA International Symposium, Trondheim, Norway.

Kalam M.Z., El Mahdi A., Negahban S., Bahamaish J.N.B., Wilson O.B., and Spearing M.C., "A Case Study to Demonstrate the Use of SCAL Data in Field Development Planning of a Middle East Carbonate Reservoir," SCA2006-18, presented at the 2006 SCA International Symposium, Trondheim, Norway.

Kalam, M. Z., Al Dayyani, T., Clark, A., Roth, S., Nardi, C., Lopez, O. and Øren, P. E., "Case study in validating capillary pressure, relative permeability and resistivity index of carbonates from X-Ray micro-tomography images", SCA2010-02 presented at the 2010 SCA International Symposium, Halifax, Canada.

Knackstedt, M.A., Arns, C.H., Limaye, A., Sakellariou, A., Senden, T.J., Sheppard, A.P., Sok, R.M., Pinczewski, W.V. and Bunn G.F., "Digital core laboratory: Reservoir-core properties derived from 3D images," *Journal of Petroleum Technology,* (2004) 56, 66-68.

Lopez, O., Mock, A., Øren, P. E., Long, H., Kalam, M. Z., Vahrenkemp, V., Gibrata, M., Serag, S., Chacko, S., Al Hosni, H., Al Hammadi M. I., and Vizamora, A., "Validation of fundamental carbonate reservoir core properties using Digital Rock Physics", SCA2012-19, SCA 2012 International Symposium Aberdeen, United Kingdom.

Lopez, O., Mock, A., Skretting, J., Petersen, E.B.Jr, Øren, P.E. and Rustad, A.B., 2010, Investigation into the reliability of predictive pore-scale modeling for siliciclastic reservoir rocks, SCA2010-41 presented at the 2010 SCA International Symposium, Halifax, Canada.

Mu, Y., Fang, O., Toelke. J., Grader, A., Dernaika, M., Kalam, M.Z., 'Drainage and imbibition capillary pressure curves of carbonate reservoir rocks by digital rock physics', SCA 2012 – Paper A069, Aberdeen, United Kingdom.

Øren, P.E. and Bakke, S., Process Based Reconstruction of Sandstones and Prediction of Transport Properties, *Transport in Porous Media*, 2002, 46, 311-343

Øren, P.E, Antonsen, F., Rueslåtten, H.G., and Bakke, S., 2002, Numerical simulations of NMR responses for improved interpretation of NMR measurements in rocks, SPE paper 77398, presented at the SPE Annual Technical Conference and Exhibition, San Antonio, Texas.

Øren, P. E., Bakke, S. and Arntzen, O. J., 1998, Extending predictive capabilities to network models, *SPE J.*, 3, 324–336.

Øren, P.E. and Bakke, S., "Process Based Reconstruction of Sandstones and Prediction of Transport Properties," *Transport in Porous Media,* (2006) 46, 311-343.

Øren, P.E., Antonsen, F., Rueslåtten, H.G., and Bakke, S., "Numerical simulations of NMR responses for improved interpretation of NMR measurements in rocks," SPE paper 77398, presented at the 2002 SPE Annual Technical Conference and Exhibition, San Antonio, Texas.

Øren, P. E., Bakke, S. and Arntzen, O. J., "Extending predictive capabilities to network models," *SPE Journal,* (1998) 3, 324–336.

Ramstad, T., Øren, P. E., and Bakke, S., 2010, Simulations of two phase flow in reservoir rocks using a Lattice Boltzmann method, *SPE J.*, SPE 124617.

Youssef, S., Bauer, D., Bekri, S., Rosenberg, E. and Vizika, O., "Towards a better understanding of multiphase flow in porous media: 3D in-situ fluid distribution imaging at the pore scale," SCA2009-17, presented at the 2009 SCA International Symposium, Noordwijk, The Netherlands.

Wu, K., Jiang Z., Couples, G. D., Van Dijke, M.I.J., Sorbie, K.S., 2007. "Reconstruction of multi-scale heterogeneous porous media and their flow prediction," SCA2007-16 presented at the 2007 SCA International Symposium, Calgary, Canada.

Wu, K., Ryazanov, A., van Dijke, M.I.J., Jiang, Z., Ma, J., Couples, G.D., and Sorbie, K.S., "Validation of methods for multi-scale pore space reconstruction and their use in prediction of flow properties of carbonate," SCA-2008-34 presented at the 2008 SCA International Symposium in Abu Dhabi, UAE.